UNLOCKING EMOTIONS IN DESIGN

A COMPREHENSIVE GUIDE TO KANSEI ENGINEERING

With worked examples for learners

Simon Schütte

Anitawati Mohd Lokman

Shirley Coleman

Lluis Marco Almagro

FSC
www.fsc.org
MIX
Papper från
ansvarsfulla källor
Paper from
responsible sources
FSC® C105338

Impressum

Author contact: simonschutte@gmx.net

Illustration book cover: Johanna Slunga and Ludvig Almquist
Proofreading: Shirley Coleman

Publisher: BoD – Books on Demand, Stockholm, Sweden
Print: BoD – Books on Demand, Norderstedt, Germany

ISBN: 978-91-8007-036-2

About the Authors

Simon Schütte is an Associate Professor at Linköping University in Sweden at the division for Product Realisation. He started research on Kansei Engineering in 2001 and has since developed a Europeanized version that can be applied in industrial affective product development processes.

Anitawati Mohd Lokman is a full Professor at Universiti Teknologi MARA in Shah Alam, Malaysia. Her research focuses on emotional user experience including Kansei Engineering. Anitawati has a long experience in application and development of Kansei Engineering techniques.

Shirley Coleman is Technical Director and Principal Research Associate at Newcastle University in the UK. Her expertise of research focuses on statistical modelling, problem solving, data visualization and business statistics. Shirley has many years' experience in the usage and application of Kansei Engineering

Lluis Marco Almagro is an Associate Professor at Polytechnical University of Catalunya in Barcelona, Spain. His field of research is industrial statistics. In 2011 he wrote his PhD Thesis on the topic of evaluation methods in Kansei Engineering technique.

Table of Content

Part 1 Introduction to Kansei Engineering

Part 2 Kansei Engineering Methodology in Practice

Part 3 Kansei Engineering Product Stories

Part 4 Exercises, small group work and ideas

PART 1

Introduction to Kansei Engineering

Kansei Engineering is a relatively new method in product development. With its origin in Japan it is based on a product realization philosophy embedded in the Japanese culture. As Kansei Engineering spreads to Western countries it is mostly seen as a tool for integrating feelings, emotions and subjective values of users, consumers and buyers of products and services.

As such it needed to be modified and altered from the traditional philosophy of Kansei Engineering into a tool for standardized product development on an international level. Part 1 explains the basic underpinnings of this tool as well as its theoretical base.

Chapter 1: A Notion of Kansei

Kansei

Kansei is a Japanese term meaning affect, emotion or feeling, and has been a major area of research in product development for over 40 years, primarily in Japan, China, and Korea. In recent years, the concept of Kansei in product development has been spreading to Europe and other parts of the world. Various groups have been established to promote Kansei, including regional organizations such as the Japanese Society of Kansei Engineering and the European Kansei Group. The bi-annual Kansei Engineering and Emotional Research (KEER) conference has grown to become an umbrella organization for research on the topic. The diverse perspectives and methods of Kansei have fostered a wider interpretation of the term and inspired the creation of various applications. The community welcomes different approaches and provides a platform for the exchange of ideas and further development. Kansei has evolved from a psychological method linking basic emotional expressions with product features to a complex collection of affective product development methods with diverse approaches, tools and applications.

The Mystique of Kansei

The term "Kansei" is of Japanese origin and roughly translates to "feel, emotion, or affect." However, its true meaning is often only understood in conjunction with a corresponding term. In contemporary dictionaries, the term Kansei is often listed alongside secondary expressions, such as "Kansei Engineering" or "Kansei Technology," but also "Kansei Poetry" or "Kansei Food." The phrase "brush up your Kansei" also exists, meaning to sharpen and improve one's attitude. This indicates that the term Kansei often adds an emotional-affective dimension to the corresponding word.

Kansei is difficult to define more precisely as it stems from a long Japanese tradition and is closely connected to Japanese culture and philosophy. This makes the term Kansei desirable for a novel dimension of research as it still holds a degree of purity and novelty, allowing researchers to define research related to affect, emotion, and personal preferences anew.

In the context of product development, the expression of Kansei describes an affective approach by product developers. Traditional product development methods use fact-based input to achieve customer satisfaction, which is mostly functional. Kansei adds an emotional aspect to product development and enables developers to cope with mostly qualitative, affective input data to create emotionally pleasing products.

However, practitioners use Kansei in different ways, due to its ambiguity. Many researchers and practitioners have found and published their own definitions to promote their personal take on the field. Although this may seem chaotic at first glance, it is actually a rich source of novel ideas and approaches and is widely seen as an advantage.

Definition of the term Kansei as used in this book

In Japanese culture, the term Kansei is mostly associated with the appreciation of beauty and the development of refined taste. Although many Kansei researchers therefore focus solely on the aesthetic aspects of artifacts, subtle and hidden technical properties also affect the emotional experience of the artifact for humans. This book establishes a bridge between the affective properties of artifacts (Kansei) and design factors (engineering), encompassing both the aesthetic and technical aspects of the artifact under the term Kansei.

The Japanese philosophy of Kansei emphasizes the use of intuition, empathy, and emotional intelligence to create products, spaces, and experiences that resonate with people on a deeper level. Kansei designers and artists aim to elicit emotional responses that provide users with a sense of pleasure, satisfaction, and well-being. They emphasize the need to create works that harmonize with nature, integrate the old with the new, and respect the cultural traditions and values of the society in which they are created.

This philosophy has deep roots in Eastern culture and has now spread around the world. However, practitioners in countries outside Japan and Eastern culture may not have an inherent understanding of the underlying philosophy. Therefore, in this book, Kansei is presented as a tool for product and service design rather than just an artistic tool, acknowledging its origins while also recognizing its global application.

Harnessing the Power of Kansei

So Kansei is a complex concept that elicits powerful feelings and emotions - yet is difficult to describe. Despite its intangibility, the Kansei is crucial for decision-making, as emotions are involved in virtually every decision humans make, according to Damaiso (1996). This means, the Kansei can - and should - be harnessed for practical purposes such as marketing, nudging technology, and product development. However, in order to use the Kansei in an industrial context, it must first be made visible and measurable, allowing people intentionally to manipulate the Kansei of objects in their environment. Although the Kansei cannot be fully measured, human reactions to emotions such as affection, desire, disgust, and inconvenience can be quantified. From there, conclusions can be drawn about the quality of the Kansei, which can be applied in the aforementioned areas. Ultimately, while the Kansei may remain an enigma, we can still utilize its impact to create more effective (and affective) products and services.

Chapter 2: Kansei Engineering

Overview

Kansei Engineering is a technology that combines Kansei with engineering fields in order to create products and services that match consumers' implicit needs and desires (Lokman, 2010). It enables the translation of implicit requirements for the affective quality of human interaction into explicit and aesthetic properties that can be designed into products (Oluwafemi & Yamanaka, 2014). This is achieved by identifying how the affective experience (such as pleasure, trust and engagement) of a product or service influences consumers' responses, and manipulating those specific properties or features of the product or service that determine the response.

Kansei Engineering was originated by Professor Mitsuo Nagamachi in Japan in the 1970s. The basic methodology of Kansei evaluation and analysis was established in the middle of the 1980s (Nagamachi & Lokman, 2010, 2015). Initially, Nagamachi used the terms "emotion" or "image technology" to represent Kansei in his methodology, but the term 'Kansei Engineering' was first used in 1986, after the basic methodology was established (Nagamachi, 1995). During the 1980s, many Japanese industries, such as the automotive and electric home appliances sectors, incorporated Kansei Engineering as an important development process.

Kansei Engineering encompasses the use of all methods that enable products or services to be designed to satisfy affective needs better.

Researchers now use physiological measurements such as electroencephalography (Yoshida et al.,) and eye tracking (Tomico et al., 2008), analyse self-report questionnaires using item response theory (Camargo & Henson, 2015), and use advanced modelling techniques such as neural network analysis and sparse Bayesian learning (Tipping, 2001). Qualitative research techniques such as empathy mapping and phenomenology are also used. These enable designers to understand how different design solutions influence the affective responses of users.

The diversity of these techniques reflects the diversity of the backgrounds of the researchers, and Kansei in the West has therefore become richly interdisciplinary, overlapping with and complementing design, engineering, psychology, affective computing, human-computer interaction, user-centred design, cultural computing, marketing and business, service design, and neuroscience. This is reflected in the research community by the use of terms such as Kansei Design and Kansei Science. Beyond the Kansei community, others pursue what we would characterize as a Kansei-like approach, even if they do not necessarily recognize the term, such as industry-based consumer researchers (Lévy, & Yamanaka, 2006).

Engineers are often interested in people's affective responses to products, and seek to create models relating perceptions to the physical properties of stimuli. Engineers are essential for understanding the physical properties of stimuli and the mechanics of human interactions with them (including the tribology of touch). However, some engineers seek linear relationships between perception and the stimuli's properties where linear relationships are unlikely to exist. One of the significant aspects of KE methodologies is analyzing the non-linear nature of Kansei and physical properties such as colour, shape and surface finish.

Types of Kansei Engineering

Product emotion has become a critical factor in determining consumer satisfaction and product success in the market. Historically, Japan has been at the forefront of innovative product development, and their success has been greatly attributed to their sensitivity to the implicit emotions of consumers, known as Kansei, which has led to the emergence of Kansei Engineering. The implementation of Kansei Engineering involves several steps that utilize tools and methods adapted from different disciplines, such as marketing, psychology, and statistics. Kansei Engineering typically employs a combination of qualitative and quantitative approaches. The literature contains numerous approaches for identifying Kansei for a target group and focus group, including self-reporting systems or ethnographic techniques, but the focus is always on the emotional aspect of the user's product experience.

In a 2010 study, Lokman provides a comprehensive overview of Kansei Engineering methodology, emphasizing the importance of emotional design and how Kansei Engineering can be utilized to create emotionally appealing products. Figure 2.1 illustrates the types of Kansei Engineering implementation utilized in the process of identifying Kansei factor certainty and determining the new design of Kansei products.

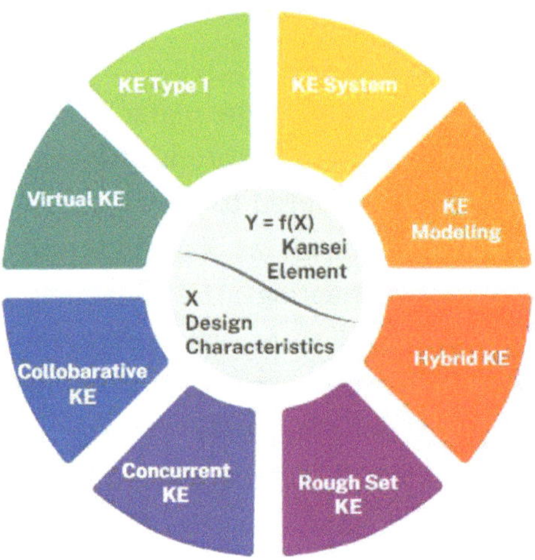

Figure 2.1: Types of Kansei Engineering (Lokman, 2010).

The implementation of Kansei Engineering has evolved over time and can be categorized into several types, as shown in Figure 2.1 and are more elaborated in Lokman (2010). These types include:

1. **KE Type I, or Category Classification:** This approach involves breaking down the core idea of a new product into its associated emotional and design elements. Qualitative research techniques like the KJ Method are used to analyze data. A famous example of this method in action is the development of the Mazda Miata sports car (compare: Nagamachi & Lokman , 2015)

2. **KE Type II, or Computer-Aided KE System (KES):** This system relies on databases and inference engines to create a computerized system that interprets consumer emotions and transforms them into design elements. Examples include

systems used for house design and fashion image development (Nagamachi & Lokman 2009).

3. **KE Type III, or KE Modelling:** Here, mathematical modeling is the foundation for a computerized system that handles fuzzy logic and machine intelligence. An example of this type is word sound diagnostic systems (Nagamachi 1999).

4. **KE Type IV, or Hybrid KE:** This approach combines Forward KES and Backward KES to create a hybrid system. This type allows for iterative processes, from initial design elements to Kansei evaluation. It's evident in studies attributed to Nagamachi & Matsubara (1997).

5. **KE Type V, or Virtual KE:** This method integrates KE techniques into virtual reality, enabling consumers to explore Kansei-driven products in a virtual environment. For instance, Matsushita Electric Works used this approach in designing kitchen cabinets (Enomoto et al., 1993).

6. **KE Type VI, or Collaborative KE:** This approach involves designers and consumers from different locations working together using a shared Kansei database to develop new product designs. An example is the Internet Collaborative Design System (Ishihara et al., 2005)

7. **KE Type VII, or Concurrent KE:** In this approach, representatives from various company departments or experts in related fields collaborate to evaluate and analyze Kansei aspects. This method provides a holistic perspective on product design, as demonstrated in research on shampoo container design (Nagamachi, 2000)

8. **KE Type VIII, or Rough Sets KE:** This type is particularly useful for dealing with unclear and uncertain Kansei data. It relies on group meanings to establish decision rules independently and is commonly applied in studies involving beer can design (Nagamachi et al., 2006) .

Each of these approaches has its own strengths and weaknesses, and the choice of approach depends on the specific needs and goals of the product development team. However, the ultimate goal of Kansei Engineering is to create emotionally appealing products that meet the needs and desires of users. The choice of KE type depends on the objectives and strategies of the organization, designers, or Kansei Engineers involved in the assessment of Kansei.

Chapter 3: Measuring the Kansei

Overview

Kansei measurement focuses on identifying and understanding users' emotional responses to products, services or situations. The use of both physiological and psychological measurements has the potential to improve the design of products or services by incorporating user emotional experiences into the design process. Figure 3.1 shows different ways to access peoples' Kansei.

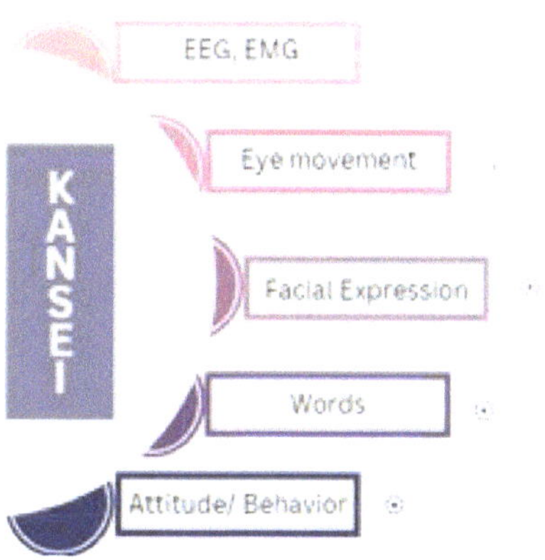

Figure 3.1: Kansei Channels (Lokman, 2010).

Several researchers have contributed to the development of Kansei Engineering, including Nagamachi, Lokman, Lévy, Schütte, Marco-Almagro and others. Nagamachi and Lokman have written a comprehensive guide to Kansei Engineering and its practical applications in product and service design (Nagamachi & Lokman,

2015). They discuss various techniques for measuring physiological and psychological responses Lévy's research has focused on developing a theoretical framework for understanding emotional responses to product designs based on cognitive and emotional psychology (Lévy, 2013).

Physiological vs Psychological Measurement

Kansei measurement is an essential process for incorporating emotion into design. Since Kansei is a subjective, vague, and unstructured concept, it cannot accurately be identified directly. Therefore, indirect assessment methods that utilize alternative expression approaches are necessary. Kansei measurements are typically divided into two categories: physiological and psychological measures. Depending on the task at hand however, the way to measure must be appropriate. While equipment and expertise for many of the physiological measurements used to be hard to come by and expensive, in recent years prices have been dropping and resources are now more readily available for future experiments. Price and availability are however just one reason for the lesser employment of physiological measurement methods. Another reason is the fact that emotions, feelings and impressions are complex and difficult to express using those methods. Hence, they are at least accompanied by psychological measurements which tend to be more sensitive to affective impressions, yet less objective. In most studies in affective product development, psychological measurements are dominant. This book reviews both types of measurement but uses psychological measurement for the practical examples.

Physiological Measurement

To accurately measure the Kansei of users, Kansei Engineering often utilizes physiological measurements that provide objective data on emotional and physical responses. One widely used method is Electroencephalography (EEG), which measures the electrical activity of the brain to provide insight into users' emotional states such as excitement, frustration, or engagement. EEG has been used in various contexts, including product design, marketing and human-computer interaction (HCI), to measure users' emotional responses (Trapsilawati, 2019; Ding et al., 2010, Mitsukura, 2020).

Another physiological measurement that has proven useful for Kansei Engineering is Galvanic Skin Response (GSR). GSR measures the electrical conductivity of the skin and can be used to determine users' levels of emotional arousal and stress (Laparra-Hernández et al., 2009). Other physiological measurements that have been used in Kansei Engineering include Heart Rate Variability (HRV), which measures the variation in time intervals between successive heartbeats, Electromyography (EMG) muscle load measurements and Eye Tracking, which measures the eye movements and gaze behaviour of users (Köhler et al., 2015).

The goal of these physiological measurements is to identify user behaviour, response, and body language, all of which can help designers create products that better align with users' emotional needs and preferences. By incorporating these physiological measurements into the Kansei Engineering process, designers can gain a deeper understanding of users' emotional responses and create products that better meet their needs (Nagamachi, 2011).

Psychological Measurements

In contrast to physiological measurement methods, psychological measurement is related to the human mental state involving user behaviour, expressions, actions, and responses. In this case Kansei measurement employs tools for designing products based on the emotional responses and subjective feelings of users. Psychological measurement tools are crucial components of this approach, as they help designers to identify and understand the emotional responses that people have to different products or design features. This can be measured using self-reporting systems.

There are several different psychological measurement techniques used in Kansei Engineering, including semantic differential (SD) scale, different emotional scale (DES), paired comparison analysis, and free labelling system. These methods are used to capture the emotional responses of users and to identify the specific design elements that influence these responses.

Semantic differential scales (SD) are an important psychological measurement. The technique involves presenting participants with bipolar adjectives, such as "good" vs. "bad", "comfortable" vs. "uncomfortable" or "stylish" vs. "unstylish" and asking them to rate the concept or object on a scale between these opposing adjectives. By analysing the resulting data, researchers can gain insight into the participants' experiences and opinions towards the concept or object being studied. The SD technique has been used in various fields, including psychology, marketing and design, to understand how individuals perceive and respond to different stimuli (Snider & Osgood, 1969). Examples are given in Figure 3.2.

Paired comparison analysis is another well-established technique. In this approach, participants are presented with pairs of designs or products and asked to choose which they prefer based on their emotional response. This approach is useful for identifying specific design elements that are responsible for emotional responses (Ramanathan et al., 2023). Free association is another important technique used in Kansei Engineering. This approach involves asking participants to provide spontaneous responses to a product or design. These responses can then be analysed to identify underlying emotional associations and inform the design requirements (Nagamachi, 2011).

As elaborated above, some of these psychological and physiological interpretations are shown through eye, skin reactions, brain waves (EEG), electromyography (EMG), heart rate and other types of measurements. Kansei researchers need to think about which input is appropriate to reach the user's Kansei and how to measure such interpretations. Kansei's access channels are often not just one, but may be a combination of several.

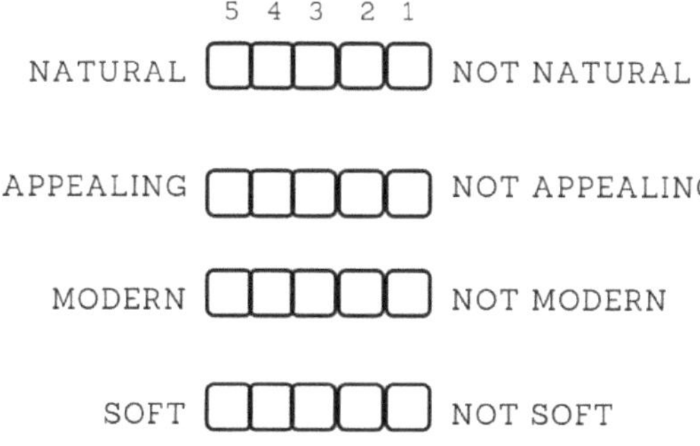

Figure 3.2: Example of a Kansei-Checklist (Lokman, 2010) .

Quantitative Measurement of the Kansei

The Quantitative approach in Kansei engineering integrates various design elements and the target user into a single development strategy. It also involves identifying a specific set of Kansei Words (KW) that match the product domain, typically requiring a set of 30 to 40 adjectives or short sentences that can be used to translate emotions. To identify KWs, the following steps can be followed: first, the company's strategy is reviewed to understand their target audience and objectives for using Kansei engineering. Next, KWs related to the design domain are identified, for example, referencing vehicle magazines to select KWs that suit the purpose of the study for vehicles, or communicating with buyers or prospective buyers for clothing design.

Once the Kansei Words (KW) have been identified, they can be processed to create a scale, such as a 5-point semantic differential (SD) scale, or a representation of human emotional response by using socio-physical and psychophysical equipment to measure the emotional responses. Then, the study proceeds to an experimental evaluation process, which typically includes 20 to 25 different types of products as study samples. The designs of the specimens are classified into distinct design elements and attributes, which specify details of the design elements that will be used in design specification. Design elements are classified into attributes (such as size, shape, and colour) and values (properties within those design attributes, such as blue, white or red for colour). The more attributes and values that can be classified, the more detailed the product design specifications can be, but then more sample specimens are needed.

The specimens can then be evaluated by specific subjects, such as employees of the company, designers, R&D personnel, or target market groups. Typically, 20 to 30 people are used as subjects in this type of Kansei research. To avoid biased or sequential effects, specimens can be shown in a random order to each subject during the assessment session. Alternatively, KWs can be presented in a random order for the same purpose. It is also important to prevent subject fatigue during the assessment process. The time required for each evaluation session can be estimated by multiplying the number of KWs by the total number of samples, and the time taken to record one response. As a rule of thumb, a person will take 2-3 seconds to record their response.

The data collected can then be analysed using multivariate techniques, which will be described in later sections. After the analysis, Kansei researchers must invite one or more domain experts or designers to interpret the data together to enable innovation and new product design. The researchers' descriptions of research findings will help the experts to understand the findings of Kansei Engineering research.

Qualitative Approach in Kansei Engineering

One qualitative approach in Kansei Engineering is called The Category Classification Method (refer to Figure 3.3). It is a highly effective and user-friendly approach that has been implemented in various Kansei Engineering research studies. This method should primarily be used to determine the main concept of a new product, which can be done through interviews or marketing surveys. For instance, Nagamachi (2010) described the study by Mazda who conducted a survey by taking photographs of young drivers using the car's horn, selecting keywords that represent each image and synthesizing them to determine the main concepts. In another example, Deesse hair shampoo and Milbon hair treatment's R&D teams interviewed women in salons to collect data on their hair problems, which was analysed using Quantification Theory Type III in Kansei Engineering to decide on the key concepts. Both Mazda and Milbon then broke down the key concepts into sub-concepts through discussion with relevant experts, continuing the process until several important sub-concepts were identified. In the case of Mazda, ergonomic experiments were also conducted to determine detailed design specifications. For instance, one experiment focused on gear stick length and identified that a 9.5 centimetre gear stick gives users the feeling and perception of control.

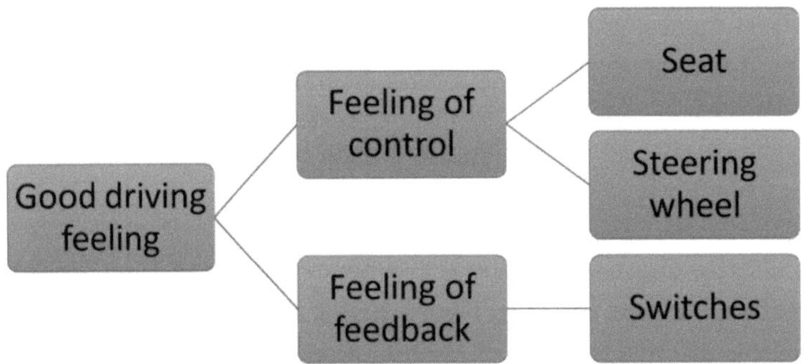

Figure 3.3: Example of Category Classification Method (compare: Nagamachi & Lokman, 2015).

In summary, the Category Classification Method is a process that shapes key concepts through qualitative-described surveys, which are then broken down into sub-concepts to generate design specifications. To achieve this, several factors must be considered, such as determining the main concept through appropriate study and analytical methods, breaking down the key concept into detailed keywords, and integrating those keywords into the new design specification.

Chapter 4:
General Kansei Engineering Model

A six-step process guide

It is helpful to have a clear process to guide Kansei research. In this book we follow the methodology identified in Figure 4.1. The rest of Part I of the book gives guidelines for stepping through the six steps involved in this methodology.

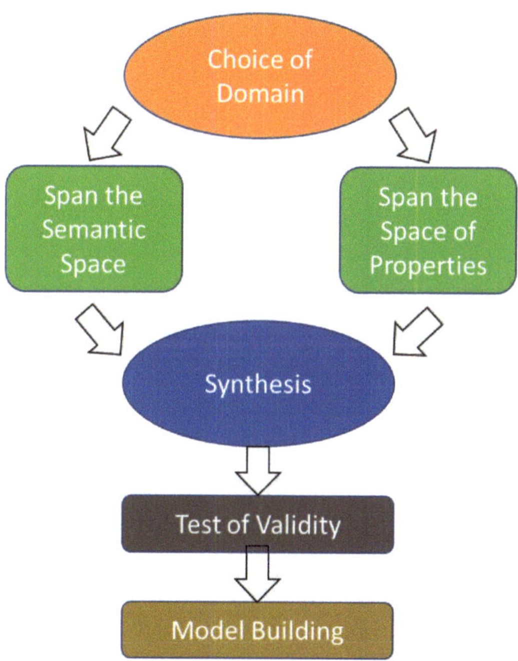

Figure 4.1: General Kansei Engineering model (Schütte et al., 2004).

1.Choosing the Domain

Selecting the domain for a new product involves identifying the target audience and market niche, as well as specifying the product's features. Based on this information, a collection of product samples is chosen to represent the domain. The Kansei Domain refers to the ideal concept behind a certain product, similar to how everyone knows what the perfect idea of a circle looks like, even though a perfectly round circle cannot physically be drawn by hand. The Kansei domain is abstract and encompasses both existing products, concepts, and potential design solutions. The goal of this first step is to define the domain and gather representative samples that cover as much of the domain as possible.

2.Defining the Semantic Space

For practical reasons spanning the Semantic Space has been subdivided into three steps as presented in Figure 4.2. Using the desired domain as a starting point, low level Kanseis also called 'Kansei Words' are collected, describing the considered product. In 'Kansei Structure Identification' higher-level Kanseis are then identified from this set using a number of possible tools. In Kansei Engineering literature these higher-level Kanseis are sometimes also referred to as 'Kansei Engineering Words'. Finally, the data is compiled in a standardised way in order to facilitate the following synthesis phase. If important Kansei Words are missed in this step, the result may have significantly limited validity. Hence, it is better to select more rather than fewer words than necessary.

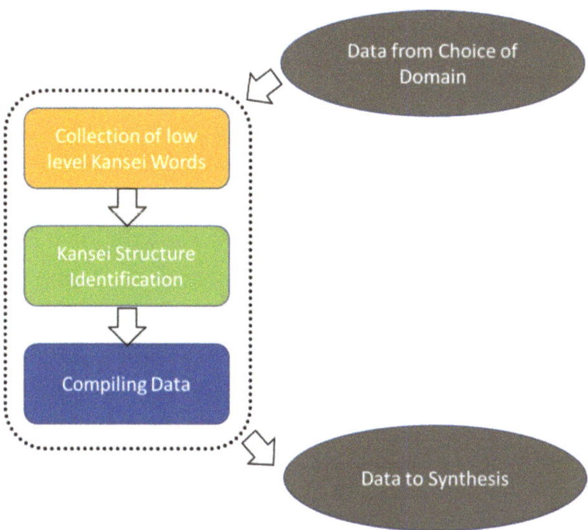

Figure 4.2: Procedure of Spanning the Semantic Space.

Assembly of Kansei Words

Kansei Words describe the product domain. Often these words are adjectives but other grammatical forms are possible. For example when describing the domain "wrist-watches", adjectives like effective, robust, quick are relevant but also verbs and nouns such as "beautiful"/ "beauty" can occur (Simon Schütte, 2005). All available sources have to be used to ensure a complete selection of words, even if the words emerging seem to be similar or the same. Suitable sources can be:

- Magazines
- Pertinent Literature
- Manuals
- Experts
- Experienced Users

- Related Kansei Studies
- Ideas, visions

An important point is also to translate ideas and visions into Kansei Words because non-existing solutions should also be considered. In this way Kansei Engineering can be used as a creative product development tool, which generates innovative solutions. The task is to describe the domain, not the existing products. Depending on the domain considered, the number of existing Kansei Words generally varies between 50 and 600 words (Nagasawa, 1997). Since it is of great importance to cover the whole Kansei, the word collection is continued until no new words occur. The data gathered will critically influence the validity of the results, especially if important words are missing.

Tools for Semantic Structure Identification

For the identification of the Kansei structure, different methods have been developed, tested and made available for use. These methods can roughly be divided into two different groups:

1. Manual Expert Methods
2. Statistical Methods

Manual Expert Methods

Expert groups often prefer manual methods. The Kansei Words are grouped and summarised according to the participant groups' preferences and needs. Tools supporting this are:

- Affinity Diagram (Bergman & Klefsjö, 2002)
- Designers' choice
- Interview techniques

The major disadvantage of manual expert methods is that experts can fail to identify the patterns and groupings in the words. An alternative is to ask the users of the products about their Kansei and what they consider to be the important Kansei. Typically, this is done by a questionnaire given to a customer group.

Statistical Methods

Using statistical methods to evaluate the gathered material quantifies the affinity between the different Kansei Words. The following statistical methods are often used:

- Principal Component Analysis (Osgood & Suci, 1969)
- Cluster Analysis (Hair et al., 1995)
- Quantification Theory Types II, III, IV (Komazawa & Hayashi, 1976)
- Neural Networks (Ishihara et al., 1995)

Each of these statistical methods aims to group the 50 to 600 Kansei Words into a smaller number of groups, typically 30 to 40, representing the key components of the Kansei structure.

3.Defining the Space of Properties

As shown in Figure 4.1. the product domain is described both from a semantic perspective and a physical perspective. Both perspectives are presented in the form of vector spaces.

However there are significant differences in the theoretical background of the two spaces. Whereas the semantic descriptions possess a well researched theoretical background based on e.g.

Osgood's Semantic Differential Technique (Osgood & Suci, 1969), there is no similar theory for the Space of Product Properties. Hence, there is no consistent way of developing the Space of Properties. At the same time, the selection of properties is essential (Nagasawa, 2002). Few studies really evaluate the affective impact and the importance of the product properties on the user. Often they are assumed to be relevant, given by the client company, or even chosen randomly. In the majority of cases however, the product properties are chosen due to the feasibility of producing product examples to present in the study (Kanda, 2002).

How can we ensure that the properties chosen are genuinely pertinent to the user or customer within the given context? What occurs if a characteristic picked for inclusion holds little significance for the user? Likewise, what if the selected properties don't cover all the necessary aspects, omitting a crucial feature? For instance, in a Kansei study, participants were tasked with evaluating the quality perception of a postal service. The samples varied in numerous attributes such as delivery time and batch tracking capability. If these critical attributes aren't considered for assessment, the final outcome will lack accuracy. Even more concerning, the absence of a vital property can go unnoticed. As a result, it becomes essential to rate the importance of various product properties and use this as a criterion for selection.

Certainly, there are methods capable of appropriately selecting product properties for Kansei Engineering. However, the challenge lies in their lack of structure and general applicability for this specific task. This section will attempt to compile methods from various studies and also incorporate techniques from related fields that serve similar objectives. Perhaps the most notable research gap pertains to establishing a systematic approach for constructing the "Space of Properties."

Collection of Kansei Words

The methodical compilation of properties for Kansei Engineering, which are essentially properties suitable for a Kansei Engineering investigation, mirrors the approach used for gathering and choosing Kansei Engineering terminology. This process can be roughly segmented into three stages, as depicted in Figure 4.3. During the collection phase, inspirational content related to a specific product domain is amassed from diverse sources, and potential properties are pinpointed. In the subsequent stage, these properties are organized based on specific criteria. The quantity of properties is then reduced by singling out the most crucial ones. Only properties with the most pronounced affective influence proceed to the subsequent stage for further assessment. Finally, example products are found possessing those properties chosen and representing in this way the space of properties. Depending on the method used for relationship identification the assembly of products can vary. In conformity with the building of the Semantic Space, the raw data is collected from different sources.

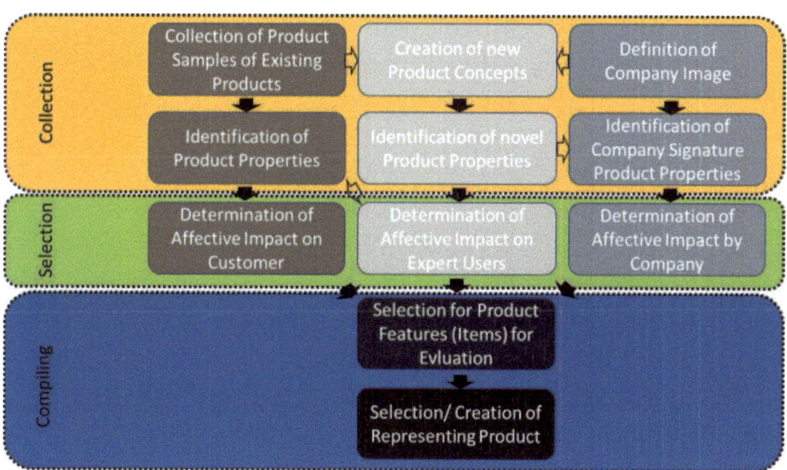

Figure 4.3: Spanning the Space of Properties (Schütte et al., 2004).

Typically, existing products provide a wide variety of potential properties, which can be integrated into new products. Getting inspiration from already existing products is one of the most common ways of identifying relevant properties. As seen in Figure 4.3, sources for collection are usually found in literature, technical datasheets, magazines etc. For identification, an assembly list of properties is often sufficient. The determination of importance and selection of properties with the highest importance and affective value is preferably done by customer representatives. To facilitate the work of gathering raw data, tools such as focus groups or one-to-one interviews can be used. For determining the importance, Pareto diagrams (Bergman & Klefsjö, 2002), for example can be useful.

In almost all Kansei Engineering studies carried out within industrial product development projects, a central specification must also fit the company image. Companies therefore tend to integrate characteristic features in their products. The right hand column in Figure 4.3 identifies that coming from the company's existing products are the product properties which are unique to the company. The relative importance of these properties is determined together with company marketing experts. Usually the number of image properties is small so that no special tool needs to be deployed to address this aspect.

The central column in Figure 4.3, however, is the integration of new product concepts. Kansei Engineering has been criticised for not being innovative, however, this part displays how creative thinking and new ideas can be integrated into Kansei Engineering as a method. As a main source the designer's mind is used. A designer can make a sketch, a mock-up or a prototype of the whole product or parts of it. Thereby s/he creates potential new properties, which are appraised and selected by an expert group.

However, Figure 4.3 also displays that these processes do not necessarily take place separately and in isolation. In contrast, they influence each other as indicated by the arrows. The designer might get inspiration from both existing products and the company's image, which is then developed into a new solution. This new concept in turn might influence the company's decision making about which product properties to select in relevant images. Also, new trends identified by the designers may influence the choice of product properties from existing products.

Finally, all selected properties are brought together into one set of product properties from which representative products are determined or mocked up to be used in the following synthesis step. Traditionally this has been done using real existing products. In cases where no products with the right combination of properties existed, renderings were used. Today AI technology can produce adequate variants of renderings improving the data collection quality and at the same time speeding up the process considerably.

4. Synthesis step

In the synthesis step the Semantic Space and the Space of Properties are linked together as displayed in Figure 4.4. Relationships are identified and established between the Kansei Words and product properties generated in the previous steps of spanning the Semantic Space (Figure 4.2) and Space of Properties (Figure 4.3). In a study made on beer cans it was found that the colour directs the expectations of the content. As such "bitter" taste is attributed to black exterior colour in combination with an oval shaped logo type (Ishihara et al., 1998)

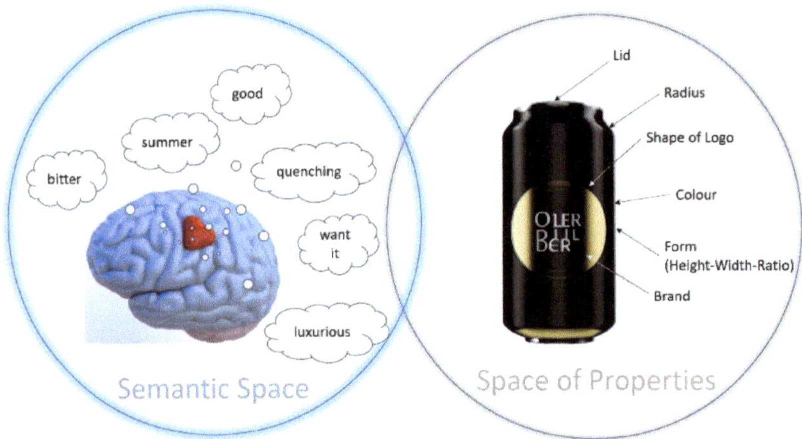

Figure 4.4: Synthesis Step.

The establishment of those links is commonly seen as a distinguishing factor of Kansei Engineering methodology. At present many different qualitative and quantitative tools are available. This book does not give an exhaustive list, but notes the methods most commonly used by the authors.

In the synthesis stage, analysis of data arising from Kansei Engineering experiments focusses on detecting which design factors affect a specific semantic scale score given to a product. As such the analysis is a form of supervised learning where the semantic scale score is the target variable and the design factors are the explanatory variables.

Relationship Identification

The tools used to identify the relationships can roughly be subdivided into five categories:

1. Manual Methods
2. Statistical Methods
3. Ranking and Rating Methods
4. Graphical Methods
5. Machine Learning

Manual Methods

Manual methods for connecting the Kansei and the different product properties are easy to perform and require comparatively small resources. These are the oldest tools and are often preferred by practitioners. One frequently used tool is referred to as Kansei Engineering Type 1 or Category identification as described by Nagamachi (1997).

Statistical Methods

As in semantic structure identification, statistical methods are used for treating great amounts of data from questionnaires. The tools used in the synthesis stage have to be modified in order to fit the requirements of Kansei Engineering. Some possible tools for statistical treatment are:

- Regression Analysis (Nagamachi et al., 2001)
- General Linear Models (Arnold, 2002)
- Quantification Theory Type I (Komazawa & Hayashi, 1976)
- Ordinal Logistic Regression and Mixed Model Logistic Regression (Marco-Almagro, 2011)

Ranking and Rating Methods

A third way of linking together the Semantic Space with the Space of Properties is semi-statistical methods, using advanced sorting mechanisms. Proven methods worth mentioning here are:

- Generic Algorithm (Tsuchiya et al., 1999)
- Fuzzy Set Theory (Shimizu & Jindo, 1995)
- Rough Set Theory (Mori, 2002)

Graphical Methods

All the above listed methods have in common that they return highly abstract results back to the user, requiring a substantial amount of expertise to interpret them. This fact almost excludes their use by industrial practitioners. In recent years, computerized graphical solution tools for mathematical problems have become accessible for common use returning results interpretable by the engineers without special training. Gephi (gephy.org) is an open source tool that has proven its suitability in recent studies. See Figure 4.5.

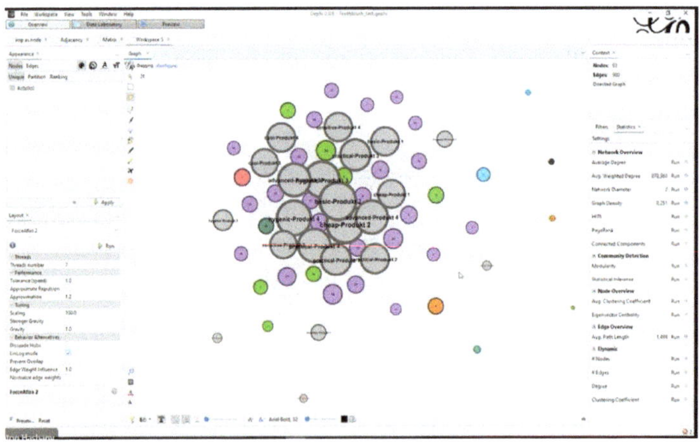

Figure 4.5 A typical output created by Gephi (gephi.org).

Machine learning

Machine learning is one of the components of data science. It is used to extract meaning from data and there are methods for identifying factors affecting a target outcome as well as general clustering of items into groups that are similar to each other in some sense. Typical methodologies include:

- Decision trees
- Random forest
- Cluster analysis

These methodologies are based on established statistical techniques such as linear models, regression, correlation and chi-square analysis. In addition to these methods, there are also deep learning algorithms which aim to detect patterns using a more black box approach that searches through all possible relationships and finds the most appropriate. Deep learning is especially effective when relationships are complex and highly non-linear. The general term for these methods is Neural Networks.

Neural networks may be of great complexity. Impressive results have been reported, in particular for pattern recognition from digitised photographs.

Neural networks are one of the techniques of Artificial Intelligence (AI). AI programs are rapidly being developed that simulate our own human intelligence and in some ways surpass it as the methods can sift through vast quantities of data evidence in a split second, much faster than humans can. AI is becoming accepted more and more. Many interesting articles can be found that relate how AI is used, for example how it can be incorporated into student assignments.

5 and 6.Model Building and Test of Validity

Finally, a mathematical or non-mathematical model is built depending on the synthesis method chosen. However, before using the model to predict future products, it has to be validated. At present, only validation methods for the Semantic Differentials are available, and there is a need for a more integrative validation concept.

The six steps described in this chapter are further illustrated in different applications in the next parts of the book.

Chapter 5:
Kansei Engineering and Data Science

Kansei Engineering is a methodology for designing products and systems that takes into account the emotional response of users. It involves identifying the sensory and emotional aspects of a product or system and designing them in a way that evokes a desired emotional response. The goal is to create products that not only meet functional requirements but also create a positive emotional experience for the user.

Data Science, on the other hand, is the process of extracting insights and knowledge from data. It involves a wide range of techniques, including machine learning, statistical analysis, data visualization, and data engineering. Data Science is used to identify patterns in data, make predictions, and inform decision making.

The idea of the intersection of Kansei Engineering and Data Science has emerged as a natural evolution of research in both fields, and it is difficult to attribute it to a single person or group. However, researchers such as Mitsuo Nagamachi, who is often credited with founding the field of Kansei Engineering in the 1970s, have been exploring the use of data analysis techniques in the field for decades.

The intersection of Kansei Engineering and Data Science involves using data analytics and machine learning to identify the emotional factors that influence product design and development. By analyzing data on user behavior and emotional responses, designers can identify which design elements are most effective at evoking specific emotional responses in users.

For example, a company may use Kansei Engineering and Data Science to design a new car that evokes a feeling of luxury and sophistication. They could use data analytics to identify which design elements are most strongly associated with these emotions, such as the materials used in the interior, the color scheme, and the shape of the car. They could then use this information to design a car that is more likely to evoke the desired emotional response in users.

Another example could be the use of Kansei Engineering and Data Science in the design of a mobile app. By analyzing data on user behavior and emotional responses, designers can identify which features of the app are most effective at engaging users and creating a positive emotional experience. They could then use this information to design an app that is more likely to be successful in the market.

In recent years, as the use of machine learning and data analytics has become more prevalent, researchers have started exploring their potential applications in Kansei Engineering. Some of the works exploring the intersection of Kansei Engineering and Data Science include Liu et al. (2023), Jin et al. (2022), and Chen et al. (2019). Chen at el. (2019) discussed the potential of combining these two fields to improve product design and development, as well as providing examples of how Kansei Engineering and Data Science have been applied in practice. Chan et al. (2018) argued that the use of data analytics and machine learning techniques in Kansei Engineering can improve the emotional satisfaction of users with products by identifying emotional factors that influence user experience and incorporating them into the product design process. Kobayashi et al. (2022) explored how deep learning techniques can be used to analyze and understand the aesthetic preferences of individuals. The authors conducted a survey asking participants to rate the aesthetic appeal of a variety of product designs, and then used deep learning algorithms to analyze the data and identify patterns in the responses. They also discuss the potential of this approach to improve the emotional response of users to products.

Other early works exploring the intersection of Kansei Engineering and Data Science include Banerjee and Chakraborty (2017), Dong and Chen (2018), and Memon et al. (2020). Banerjee and Chakraborty (2017) discussed the potential of combining these two fields to improve product design and development, as well as providing examples of how Kansei Engineering and Data Science have been applied in practice. Dong and Chen (2018) argue that the use of data analytics and machine learning techniques in Kansei Engineering can improve the emotional satisfaction of users with products by identifying emotional factors that influence user experience and incorporating them into the product design process. Memon et al. (2020) discussed the potential of Kansei Engineering and Machine Learning to work together to improve the emotional response of users to products. Their work covers a range of applications of Kansei Engineering and Machine Learning, such as product design, sentiment analysis, and image recognition.

Encouragingly, studies involving the intersection of Kansei Engineering and Data Science have been evolving. In 2019, Lee et al. proposed a method for predicting user emotions using Kansei Engineering and Machine Learning. The authors describe a model that takes into account various sensory and emotional factors to predict a user's emotional response. The paper presents a detailed description of the proposed method and evaluates its performance using real-world data. The results suggest that the proposed method can accurately predict user emotions in various contexts. Zhang et al. (2020) proposed a hybrid approach for emotion recognition in images that combines Kansei Engineering and Deep Learning. The results suggest that the proposed method can accurately classify emotions in images. Alavi and Soni (2022) proposed a data-driven approach to Kansei Engineering for product design. The paper presents a detailed description of the proposed method and provides a case study to illustrate its application.

Overall, the intersection of Kansei Engineering and Data Science is an exciting area of research that has the potential to improve product design and development by creating products that not only meet functional requirements but also create a positive emotional experience for users.

Chapter 6: Summary

Kansei Engineering is a design approach focused on creating products and services that satisfy customers' emotional needs and preferences. It involves identifying customers' emotional responses to a product, categorizing these responses into specific emotions or "Kansei Words," integrating them into the design process, and verifying their emotional impact with customers. By understanding and incorporating emotional qualities into the design process, designers can create products that better meet customers' needs and preferences.

Part 1 of the book includes a six-step process guide for carrying out a Kansei study. The general model starts with the definition of the Product or Service domain which is described from an affective semantic perspective as well as a feature perspective that comprises design parameters for the future product or service. Both spaces are interlinked in the synthesis step defining how a certain affective profile for a future product can be achieved. In the final step the conclusions are formalized in a model and verified ensuring generality.

Part 2 of the book describes Kansei Engineering methodology in practice. Part 3 offers detailed case studies in a range of application areas. Finally, Part 4 considers the pedagogic aspects of Kansei Engineering with ideas for group work and projects. It includes examples of posters created by post-graduate students who have applied Kansei Engineering in their dissertation studies.

References for Part 1

Arnold, K. (2002). Towards increased customer satisfaction. In *IKP/KMT*. Linköping University.

Bergman, B., & Klefsjö, B. (2002). *Kvalitet i alla led*. Studentlitteratur.

Camargo, F. R., & Henson, B. (2015). Beyond usability: designing for consumers' product experience using the Rasch model. *Journal of Engineering Design*, *26*(4–6), 121–139. https://doi.org/10.1080/09544828.2015.1034254

Damaiso, A. R. (1996). *Descartes' error: emotion, reason and the human brain*. Papermac.

Deogun, J. S., Raghavan, V. V, Sarkar, A., & Sever, H. (1997). Data Mining: Trends in Reserach and Development. In T. Y. Lin & N. Cerone (Eds.), *Rough Sets and Data Mining*. Kluwer Academic Publishers.

Enomoto, M., Nagamachi, M., Nomura, J., Sawada, K. (1993). Virtual Kitchen System Using Kansei Engineering. Proceedings of the International Conference on Human Computer Interaction. 657-662.

Hair, J. F., Anderson, R. E., Tatham, R. L., & Black, W. C. (1995). *Multivariate data analysis with readings*. Prentice-Hall.

Ishihara, S, Ishihara, K., Nagamachi, M., & Matsubara, Y. (1995). arboART: ART based hierachical clustering and its application to questionnaire data analysis. *IEEE Conference on Neural Networks*.

Ishihara, Shigekazu, Ishihara, K., & Nagamachi, M. (1998). Hierarchical Kansei analysis of beer can using neural network. *Proceedings of Human Factors in Organizational Design and Management - VI*, 421–425.

Ishihara, I., Nishino, T., Matsubara, Y., Tsuchiya, T., Kanda, F., Inoue, K. (2005). Kansei And Product Development (In Japanese), Ed. M. Nagamachi. Vol. 1. Tokyo: Kaibundo.

Köhler, M., Falk, B., & Schmitt, R. (2015). Applying Eye-Tracking in Kansei Engineering Method for Design Evaluations in Product Development. *International Journal of Affective Engineering*, *14*(3), 241–251. https://doi.org/10.5057/ijae.ijae-d-15-00016

Kanda, T. (2002). A Method to evaluate Human Meal Kansei. *Internaional- International Journal of Kansei*, *3*(Kanda, T.), 13–20.

Komazawa, T., & Hayashi, C. (1976). *A statistical method for quantification of categorical data and its applications to medical science* (F. T. de Dombal & F. Gremy (eds.)). North-Holland Publishing Company.

Laparra-Hernández, J., Belda-Lois, J. M., Medina, E., Campos, N., & Poveda, R. (2009). EMG and GSR signals for evaluating user's perception of different types of ceramic flooring. *International Journal of Industrial Ergonomics*, *39*(2), 326–332. https://doi.org/10.1016/j.ergon.2008.02.011

Lévy, P, Yamanaka, T. (2006). Kansei information approach for an interdisciplinary design method proposal based on intuition. *9th International Design Conference DESIGN 2006*, 1475–1482.

Lévy, P. (2013). Beyond Kansei Engineering: The emancipation of Kansei design. *International Journal of Design*, *7*(2), 83–94.

Lokman, A. M. (2010). *DESIGN & EMOTION: The Kansei Enginering Methodology*. Malaysian Journal of Computing *1*(1), 1–14.

Lokman, A. M., & Noor, N. L. M. (2006). Kansei Engineering Concept in E-Commerce Website. *Proceedings of the International Conference on Kansei Engineering and Intelligent Systems 2006 (KEIS '06)., 2006*, 117–124.

Marco-Almagro, L. (2011). *Statisical Methods in Kansei Engineering Studies*. UPC Barcelona Tech.

Matsubara, Y., Nagamachi, M. (1997a). Kansei Engineering Approach for Landscape Evaluation.Proceedings of the 13th Triennial Congress of the International Ergonomics Association. Vol. 2, 223-225.

Mori, N. (2002). Rough set approach to product design solution for the purposed "Kansei." *The Science of Design Bulletin of the Japanese Society of Kansei Engineering*, *48*(9), 85–94.

Nagamachi, M., Lokman, A. M. (1995). *Kansei Innovation: Practical Design Applications for Product and Service Development*. Taylor & Francis Group: CRC PRess.

Nagamachi, M. (2000). Application of Kansei Engineering and Concurrent Engineering to a Cosmetic Product. Proceedings of the ERGON-AXIA 2000, Warsaw

Nagamachi, M, Nishino, T., & Ishihara, S. (2001). Structural Analysis on Relations between Kansei Words. In M. G. Helander, H. M. Khalid, & M. P. Tham (Eds.), *The International Conference on Affective Human Factors Design*. Asean Academic Press, London.

Nagamachi, Mitsuo. (1995). Kansei Engineering: A new ergonomic consumer-oriented technology for product development. *International Journal of Industrial Ergonomics*, *15*, 3–11.

Nagamachi, Mitsuo. (1997). Kansei Engineering: The framework and methods. In M Nagamachi (Ed.), *Kansei Engineering 1*. Kaibundo Publishing Co., Ltd.

Nagamachi, M. (1999). Kansei Engineering: A New Consumer-Oriented Technology For Product Development. In W. Karwowski and W. S. Marras (Eds.), The Occupational Ergonomics Handbook,CRC Press, Chap. 102, 1835-1848.

Nagamachi, M., Okazaki, Y., Ishikawa, M. (2006). Kansei Engineering and Application of The Rough Sets Model. In Proceedings of IMechE 2006 (pp. 763-768), Vol. 220, Part I: Journal of Systems and Control Engineering.

Nagamachi, Mitsuo. (2011). *Kansei/Affective Engineering* (Mitsuo Nagamachi (ed.); 1st ed.). CRC Press.

Nagamachi, Mitsuo, & Lokman, A. M. (2015). *Kansei Innovation: Practical Design Applications for Product and Service Development*. Taylor & Francis Group: CRC Press.

Nagamachi, Mitsuo, & Mohd Lokman, A. (2015). *Kansei Innovation*. CRC Press LLC.

Nagasawa, S. (1997). Kansei evaluation using fuzzy structural modelling. In M Nagamachi (Ed.), *Kansei Engineering 1* (pp. 119–125). Kaibundo Publishing Co., Ltd.

Nagasawa, Shin'ya. (2002). Kansei and business. *Kansei Engineering International- International Journal of Kansei Engineering*, *3*(3), 2–12.

Oluwafemi, S. A., & Yamanaka, T. (2014). Kansei as a Function of Aesthetic Experience in Product Design. In *Industrial Applicaions of Affecctive Engineering* (pp. 83–95). Springer.

Osgood, C. E., & Suci, G. J. (1969). Factor analysis of meaning. In C. E. Osgood & J. G. Snider (Eds.), *Semantic differential technique - a source book* (pp. 42–55). Aldine Publishing Company.

Schütte, S., Eklund, J., Axelsson, J., & Nagamachi, M. (2004). Concepts, methods and tools in Kansei Engineering. *Theoretical Issues in Ergonomics Science*, *5*(3). https://doi.org/10.1080/14639220210000049980

Schütte, Simon. (2005). Engineering emotional values in product design- Kansei Engineering in development. In *Institution of Technology*. Linköping University.

Shimizu, Y., & Jindo, T. (1995). A fuzzy logic analysis method for evaluating human sensitivities. *International Journal of Industrial Ergonomics*, *15*, 39–47.

Shinya Nagasawa. (2002). Improvement of the Scheffe's Method for Paired Comparisons. *Kansei Engineering International*, *3*(3), 47–56. https://www.jstage.jst.go.jp/article/kei1999/3/3/3_3_47/_pdf/-char/ja

Snider, J. E., & Osgood, C. E. (1969). *Semantic Diffential Technique - A Source Book* (Issues 68-19874 (Libary of Congress Catalog)). Aldine Publishing Company.

Tipping, M. E. (2001). Sparse Bayesian learning and the relevance vector machine. *Journal of Machine Learning Research*, *1*(Jun), 211–244.

Tomico, O., Mizutani, N., Lévy, P., Yokoi, T., Cho, Y., Yamanaka, T., & others. (2008). Kansei physiological measurements and constructivist psychological explorations for approaching user subjective experience. *DS 48: Proceedings DESIGN 2008, the 10th International Design Conference, Dubrovnik, Croatia*, 529–536.

Tsuchiya, T., Ishihara, S., Matshbara, Y., Nishino, T., & Nagamachi, M. (1999). A Method for Learning Decision Tree using Genetic Algorthm and its Application to Kansei Engineering System. *IEEE SMC'99*.

Zhou, Z., Cheng, J., Wei, W., & Lee, L. (2021). Validation of evaluation model and evaluation indicators comprised Kansei Engineering and eye movement with EEG: an example of medical nursing bed. *Microsystem Technologies*, *27*(4), 1317–1333. https://doi.org/10.1007/s00542-018-4235-1

PART 2

Kansei Engineering Methodology in Practice

In this part the authors illustrate the working principle of Kansei Engineering methodology as presented in Part 1. As an example generated data for a fruit drink are used as it can be found in (Marco-Almagro, 2011). In this sample study several common tools and methods are put to work. There are however more methods in existence of which several are referred to in part 1.

Chapter 7:
A scientific approach to Kansei Project management

The working principles of Kansei Engineering are described in Part I and six-step guidelines are given for carrying out Kansei research. A project involving Kansei research, like any other project, will benefit from an underlying scientific approach to project management.

Here we consider two commonly used approaches: Six Sigma and Quality Function Deployment (QFD).

Six Sigma

The *Six Sigma* approach to project management is often referred to by the acronym *DMAIC* corresponding to the first five key stages of Six Sigma. These stages may be augmented by two further stages and in total the seven stages are described below.

Define (D) is where all aspects of the research question are defined including the people involved, the time window of interest, the time and resources available for the study and the quantified objectives.

Measure (M) is where all aspects of the measurements being made and used in the research project are checked for meaning, reliability, reproducibility, representativeness, error and bias. In this stage, much use is made of graphical analysis. Measurements may be plotted in time order to highlight patterns such as periodicity and autocorrelation; as histograms showing the frequency distribution of their values and as scatterplots showing the relationships between pairs of measurement sets, such as Kansei feelings for two different Kansei words or between age and a Kansei sentiment. Outliers, errors and bias are all likely to show up in the plots and can help the researcher to refine the data collection technique and to prepare the data for further analysis.

Analyse (A) is where relationships between measurements are explored and any parallels from other research projects can be investigated.

Improve (I) is where consideration is given to whether the process or product can be improved. Improvement activities are carried out, for example a statistically designed experiment.

Control (C) is where consideration is given to how any benefits from the Kansei research project can be maintained and the process or product can be continuously improved.

The two extra stages are:

Transfer (T) in which consideration is given to whether the learning from the current research project can be applied to other similar projects and situations.

Evaluate (E) in which the monetary value is calculated for any improvements brought about as a result of the project.

Quality Function Deployment

Quality Function Deployment (QFD) and Kansei Engineering (KE) are both popular methods in product design and development. The integration of QFD and KE has been discussed in the literature as a way to improve the quality of the final product. QFD provides a way to compare different products based on specific customer needs while KE focuses on linking customers' emotions and senses to the properties of a product (Mazur, 2000).

The House of Quality, which is the basic design tool of QFD, was originated in 1972 at Mitsubishi's Kobe shipyard site. It is a tool used to map customer requirements to technical specifications. It involves identifying customer needs, prioritising them, and then developing technical requirements to meet those needs. By incorporating Kansei Engineering studies for different competing products, a benchmarking profile can be developed and integrated into the House of Quality (Mazur, 2000; Lokman et al. 2016).

Integrating QFD and Kansei Engineering allows designers to understand customer needs better and develop products that meet those needs. The House of Quality provides a way to organize and prioritize customer requirements, while Kansei Engineering provides insights into customers' emotional responses to a product's design and features. By combining these two approaches, designers can develop products that not only meet functional requirements but also evoke positive emotional responses from customers.

Data issues

Both quantitative and qualitative analyses depend on data. The data may be in the form of measurement (scale) data, ordinal or nominal data. The data is the foundation stone and must be reliable and of high quality.

Kansei research often results in a large quantity of data being amassed. Care has to be taken with the recording and storing of the data; legacy issues of collection and storage methods; version control of datasets and access permissions.

There are some measurement issues which are more pertinent to Kansei research. For example, the precision of 1 to 5 ordinal scores may be suitable in some studies but in others a scoring system from 0 to 100 may be preferred.

Datasets collected during Kansei research can be represented in a number of ways but in all cases there are at least 3 dimensions. These dimensions are: people, Kansei measures and product factors. Typically each data item has three metadata: the subject who gave the measurement or opinion; the Kansei word that the feeling relates to, and the product from which the feeling arises. These three metadata are potential pivots and need to be accessible from the database in which the data is stored to enable slicing and dicing of the data. For example, a slice through the data may consist of all the responses given by a particular person or subject. This slice can be further diced to give the responses by that person for a particular Kansei word across all products, or for a particular product.

In addition each product has particular values of the selected product attributes. For example, product number 1 in the fruit juice study discussed below, is an orange coloured fruit juice drink in a glass with no straw, ice or decoration, whereas product number 12 is a yellow coloured fruit juice drink in a goblet with a straw and decoration but no ice. Each of these product attributes needs to be accessible, so that the emotional effect, for example, of the type of container can be assessed.

Database structures

Often the data are stored in Excel spreadsheets. The three data dimensions may appear in any of rows, columns and sheets.

Table 7.1 gives the database structure from 4 of the studies described in Part 3 of this book.

Table 7.1: Examples of databases.

Database	Rows	Columns	Sheets	Comments
Avatar	Kansei words (32)	Products (27)	People (17)	The products have many different attributes, some very unbalanced in the product sample, e.g. only 2 of the 27 product specimens have the attribute "robotic".
Juices	Products (16)	People (24)	Kansei words (7)	Data are presented with 2 replications. There are 5 product attributes.
Tooth-brushes	People (35)	Products (4) x Kansei words (7)	single sheet	Data are on a single sheet with products and words combined. There are 2 product attributes and also 2 people attributes.
Websites	Kansei words (40)	Products (35)	People (30)	The products have many different identifiable attributes (~260), some very unbalanced.

The numbers of cases are given in brackets, for example the Avatar database has 32 Kansei words, 27 product specimens and responses from 17 people.

Chapter 8: A Kansei study of fruit juices – As an example

By Lluis Marco- Almagro

This chapter shows an example of a Kansei Engineering study with fruit juices. This study was originally carried out as a master's degree thesis, and later included in a doctoral thesis (Marco-Almagro, 2011). The study is very simple, and was done with the purpose of having good quality data for research and teaching (we were more interested in the process than in the result, although, as you will see, the conclusions have some interest).

Choice of domain

Selecting the domain constitutes the initial phase within the proposed model for conducting Kansei Engineering (KE) studies. The process of domain selection involves not only choosing the focal product for investigation but also encompasses several additional tasks. Alongside product selection, determining the domain involves:

1. Identifying the specific target audience that the product aims to reach (Schütte et al., 2004). Participants in the study must possess characteristics matching those of the target group, essentially forming a sample from the broader population represented by the target audience. Typically, a relatively homogeneous group of individuals sharing similar personal

and socioeconomic traits is chosen. While this theoretically ensures similarity in participant ratings, practical implementation can sometimes deviate from this principle.

2. Specifying the manner in which the product is presented. This choice primarily revolves around two options: showcasing actual physical products, such as prototypes or fully functional items, or presenting representations of the product (2D depictions like photographs or 3D renditions, potentially through virtual reality setups). The number of senses engaged in processing product presentations significantly influences the perceived Kansei, referred to as the "affective channel width" (Picard, 2000). For instance, when examining perfume bottles, the affective channel is broader when participants interact with a tangible perfume bottle (enabling sight, touch, and smell) compared to encountering a photograph of the bottle. This concept is termed "proximity of presentation" by Schütte (Schütte, 2005, pp. 67-68).

3. Determining the prior experience with the product, the level of interest in the product, and the degree of interaction permitted. These factors collectively shape what is termed "proximity of interaction" (Eklund, Kiviloog, 2003), contributing to the manner in which Kansei is conveyed.

4. Defining the context for the presentation. The ambiance of the experimental environment can influence the emotional responses evoked by the product. Identical products may evoke different emotions based on the surroundings. Generally, it is advisable to select a neutral environment for conducting experiments, thereby minimizing the impact of the environment's influence as much as possible.

In our experiment, the chosen product for the KE study is fruit juices, and specifically their presentation just before being drunk. As we were interested in the presentation of the juices, the products were presented as photographs to the participants.

The participants in the study were 24 people, who collaborated with no economic compensation. There were 13 women and 11 men, with ages ranging from 17 to 59, and different educational levels (from high school to post-graduate studies).

Spanning the Semantic Space

Spanning the semantic space means choosing which Kansei words are going to be used in the study. Usually, Kansei words are adjectives. However, verbs or nouns can also be used (Schütte et al. 2004). Sometimes, even sentences can be used as Kansei words if they better describe the desired ideas.

Spanning the Semantic Space comprises three steps:

1. Setting an initial list of Kansei words.
2. Reducing the initial list of Kansei words using qualitative or quantitative methods (or a combination of both).
3. Proposing the final reduced list of Kansei words (the ones that will be used later when collecting data).

One distinguishing aspect of Kansei Engineering involves the effort to comprehensively characterize the full spectrum of emotions that a product can evoke. This endeavor's success hinges on the Semantic Space encompassing all potential emotions elicited by the product. Thus, the initial step in mapping out this Semantic Space revolves around creating an extensive inventory of Kansei words drawn from various accessible sources. The greater the length of this list, the more effective the outcome.

These Kansei words can be sourced from a variety of outlets including magazines, manuals, catalogs, websites, user input, expert opinions, or related Knowledge Engineering (KE) studies. If feasible, imaginative concepts should be translated into Kansei words to leverage KE as a tool for innovative product development (Schütte et al., 2004). The size of this initial Kansei word collection can vary widely, ranging from dozens to hundreds. This initial word compilation process should persist until no new words emerge, ensuring thorough coverage of the entire Semantic Space.

The next step involves an initial culling of the Kansei words. For instance, terms linked with material properties (e.g., metallic) or specialized jargon (e.g., art deco-like) can be directly excluded from the starting list (Jindo, Hirasago & Nagamachi, 1995).

In the context of our juice-related experiment, the preliminary list of Kansei words was created by scouring literature and online sources focusing on fruit juices. This list comprised 134 words. Eleven words were promptly removed from consideration due to their perceived connection with physical attributes of the juices rather than the sensations they evoke (e.g., decorated, with alcohol), overly generic terms (e.g., nice, ugly), or words that were challenging to associate with juices (e.g., deep, ideal). The decision was made to retain antonyms, given the difficulty in reaching consensus on which pairs should be eliminated. The initial list of Kansei words is presented in Table 8.2.

Table 8.2: Original list of Kansei words.

acidic	delicate	juicy	savory
amusing	delicious	youthful	seductive
anti-stressing	dense	laxative	sensual
anti-fatigue	detoxifying	light	silky
antioxidant	digestive	luxurious	simple
aphrodisiac	diuretic	mature	slimming
appealing	divine	mawkish	soft
appetizer	dry	mediocre	sophisticated
aromatic	easy to digest	moisturizing	sour
artificial	energetic	natural	spicy
astringent	excellent	nice	stimulating
attractive	exciting	nutritional	strengthening
bad	exotic	passionate	strong
balanced	explosive	pleasant	substitutive
beneficial	exquisite	popular	subtle taste
bitter	fabulous	powerful	sugary
bracing	fashionable	preventative	summery
brilliant	festive	protein-rich	Sweet
calming	fibrous	pure	Sylvan
Caribbean	flamboyant	purifying	Tasty
Christmas spirit	flowery	refined	Tempting
classical	foamy	refreshing	Tropical
cold	fresh	regenerative	Ugly
colourful	futuristic	reinforcing	Velvety
combinable	gluttonous	relaxing	Versatile
comfortable	good	remineralising	Vibrant
concentrated	gratifying	renovative	Vital
consistent	healthy	restorative	vitamin-rich
corpulent[1]	home-made	restorer	Warm
creamy	ideal	revitalizing	Wild
curative	infantile	rich in iron	with alcohol
dainty	intense	romantic	erotic
decorated	invigorating	salutary	
deep	irresistible	satiable	

The initial list of Kansei words must be streamlined to a manageable quantity suitable for data collection. This reduction inevitably involves a loss of information; however, the intention is to eliminate words conveying highly similar emotions, which are essentially redundant. There are various strategies for conducting this Kansei word reduction:

1. **Qualitative Approach**: Similar-meaning Kansei words can be grouped together using an affinity diagram. Within each cluster, one Kansei word is chosen to represent the entire group. Given the potentially large number of words in the initial Semantic Space, this grouping may occur across multiple stages. Since no quantitative data are being gathered, further analysis is unnecessary.

2. **Quantitative Approach**: This method requires a small dataset, and involves creating word groups by analyzing the data using multivariate techniques. Typically, a limited set of products is chosen, and individuals are asked to rate Kansei words using the same scale intended for the main data collection phase. The specific products used for this preliminary rating don't need to overlap with those used later in the data collection. These products should differ substantially to capture the full range of variations in Kansei words.

Once data is collected, averages are calculated across all raters for each product-Kansei word pair. This produces a numerical value for each Kansei word associated with each product. Using cluster analysis, Kansei words are clustered based on their ratings profiles across products. Words that closely align with others (in the sense of sharing similar rating patterns across all products) are grouped together.

3. **Combined Qualitative and Quantitative Approach**: An affinity diagram can be employed initially to group Kansei words qualitatively. This process can be repeated if necessary. Subsequently, a cluster analysis is conducted to finalize the selection of Kansei words. In cases where non-hierarchical clustering methods, like k-means clustering, are used, the clusters resulting from the affinity diagram can serve as starting points for the clustering process. This method refines the word reduction using collected data.

The last approach, which combines qualitative and quantitative elements, was employed in the fruit juice experiment. To condense the initial array of Kansei words into a smaller set, an affinity diagram was employed as the initial step.

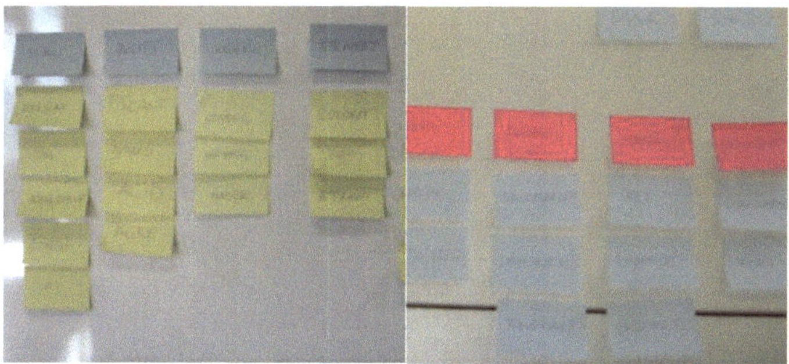

Figure 8.1: Sticky notes used for preparing the affinity diagram (first step, left; second step, right).

Each of the 123 words was inscribed onto a yellow post-it note. These adhesive notes were affixed to the wall, clustering words that exhibited similar meanings together (as depicted in the right side of Figure 8.1). This process yielded a total of thirty-three distinct groups. Within each group, the most illustrative word was selected as a label, which was then transcribed onto a blue post-it note. This procedure was repeated for the blue post-it notes to further condense the word count, resulting in a chosen representative word for each group, noted on pink post-it notes (as shown in the right side of Figure 8.1). Through this sequence, the original extensive range of words was ultimately streamlined to a mere 14 words. For a comprehensive overview of this affinity diagram process, refer to Table 8.3..

Table 8.3: Kansei words classification after the affinity diagram.

			artificial	artificial	artificial
			divine	divine	
		festive	Christmas spirit	festive	festive
			warm	warm	
reinforcing	energetic	protein-rich	stimulating	energetic	energetic
invigorative	powerful	revitalising	strengthening		
		restorative	vitamin-rich	nutritive	
		nutritive	satiable		
relaxing	calming	anti-fatigue	anti-stressing	relaxing	relaxing
erotic	exciting	romantic	aphrodisiac	seductive	seductive
sensual	attractive	tempting	explosive		
	seductive	wild	passionate		
	flamboyant	colorful	appealing	appealing	
	fresh	cold	refreshing	refreshing	refreshing
	sweet	sugary	gluttonous	sweet	juvenile
	infantile	juvenile	mature	juvenile	
	vibrant	amusing	gratifying	amusing	
dense	creamy	light	juicy	lightly	lightly
delicate	soft	silky	velvety	soft	
			refined		
		simple	comfortable	simple	
	substitutive	versatile	combinable	combinable	classical
	home-made	popular	classical	classical	
sophisticated	futuristic	luxurious	fashionable	sophisticated	
			dainty		
	Caribbean	exotic	tropical	exotic	exotic
			summery	summery	
healthy	natural	pure	curative	healthy	healthy
detoxifying	antioxidant	moisturizing	beneficial		
	salutary	preventative	vital		
slimming	laxative	digestive	renovative	purifying	

diuretic	fibrous	astringent	purifying		
restorer	remineralizing	regenerative	bracing	bracing	
			balanced	balanced	
	excellent	brilliant	fabulous	good	tasty
	bad	good	mediocre		
	subtle taste	savory	tasty	tasty	
exquisite	irresistible	pleasant	delicious	exquisite	
concentrated	intense	corpulent	consistent	concentrated	concentrated
		dry	strong	strong	
		aromatic	mawkish	aromatic	aromatic
		sylvan	flowery	sylvan	
spicy	acidic	sour	bitter	acidic	

The second phase of the reduction process involved gathering information. Four pictures depicting various juices were chosen. These selections aimed to encompass different juice styles, including distinct colors and glassware variations, such as glasses and goblets. Six individuals, comprising three men and three women from our research group, were chosen. They were then requested to assess each of the four juices using the 33 Kansei words derived from the initial affinity diagram (highlighted in blue within Table 8.3). The evaluations were conducted using 7-point Likert scales, which were also employed in subsequent data collection stages. Notably, these six participants were not involved in subsequent phases of the research.

Figure 8.2: Samples for cluster analysis, Source: Pexel.com Photographer: Engin Akyurt.

As 6 people were used, each juice had 6 ratings for each Kansei word. These 6 ratings were summarized using the average. Table *8.4* (right) shows the average points each juice got for each Kansei word.

Hence a k-means clustering was done on the 33 Kansei words, assigning each word to one of the initial 14 clusters obtained in the last affinity diagram. Table 8.4 (right) shows the 33 words in a column, each word having a number from 1 to 14 giving the initial assignation to a cluster. After the k-means procedure, only the cluster containing the word "relaxing" is kept without changes. All other clusters are modified a little bit, some become smaller, some become bigger and some keep the same size but change one of the Kansei words.

Table 8.4: Average ratings for the 4 initial juices (left) and groups before and after the k-means cluster analysis (right).

acidic	3,50	2,17	4,33	4,00
amusing	4,67	2,67	4,17	4,33
appealing	5,33	4,33	4,17	4,67
aromatic	4,17	3,50	4,17	3,50
artificial	4,33	3,50	3,50	3,33
balanced	3,50	4,83	4,50	4,67
bracing	3,33	4,50	4,00	4,67
classical	2,50	3,50	3,17	4,50
combinable	4,33	3,83	2,83	3,83
concentrated	4,17	4,50	4,00	4,00
divine	3,67	2,33	2,83	2,67
energetic	3,67	3,33	4,67	4,33
exotic	5,00	5,50	4,50	2,67
exquisite	3,67	3,83	3,50	4,17
festive	5,83	3,83	5,00	4,00
good	4,33	4,00	4,33	5,67
healthy	3,83	4,50	5,50	5,50
juvenile	4,17	3,83	3,83	4,50
light	3,83	3,33	4,17	5,33
nutritive	3,83	4,83	5,50	5,00
purifying	3,67	3,33	4,17	4,50
refreshing	5,00	3,83	4,33	5,00
relaxing	5,00	4,33	3,50	4,67
seductive	4,50	3,33	4,50	3,83
simple	2,83	3,00	4,00	5,50
soft	2,83	4,17	4,17	5,00

Before		After	
artificial	1	1	divine
divine	1	1	warm
festive	2	2	artificial
warm	2	3	nutritional
energetic	3	3	healthy
nutritive	3	4	relaxing
relaxing	4	6	appealing
appealing	5	6	refreshing
seductive	5	6	summery
refreshing	6	7	energetic
amusing	7	7	amusing
juvenile	7	7	juvenile
sweet	7	7	purifying
light	8	8	light
simple	8	8	simple
soft	8	8	soft
classical	9	8	classical
combinable	9	9	sweet
sophisticated	9	9	combinable
exotic	10	9	exquisite
summery	10	9	concentrated
balanced	11	10	festive
bracing	11	10	exotic
healthy	11	11	balanced
purifying	11	11	bracing
exquisite	12	12	good

sophisticated	4,50	3,83	4,00	3,33	good	12	12	tasty
strong	4,50	3,17	4,00	3,83	tasty	12	13	seductive
summery	5,33	3,83	5,00	5,67	concentrated	13	13	sophisticated
sweet	3,83	4,67	3,50	3,83	strong	13	13	strong
sylvan	3,50	2,50	3,83	2,83	acidic	14	13	aromatic
tasty	4,33	5,00	4,67	5,33	aromatic	14	14	acidic
warm	2,83	2,17	2,50	2,33	sylvan	14	14	sylvan

Selecting a single word from each of the final clusters presented in the right section of Table 8.3 is necessary to label the clusters. This task of choosing the most suitable word that effectively encapsulates the ideas conveyed by each cluster can sometimes be challenging.

The Semantic Space becomes fully established when the list of Kansei words to be employed in the data collection phase is clear and finalized. Occasionally, those responsible for the product may have a specific interest in determining if a particular emotion is evoked by the product, prompting them to include that word in the list. Alternatively, the individuals overseeing the experience might decide that the list of words is overly lengthy. In such situations, words can be added or removed, always with the overarching goal of ensuring that the ultimate collection of Kansei words comprehensively covers the entire semantic spectrum.

In our specific experiment, we ultimately opted for seven Kansei words: "relaxing," "artificial," "refreshing," "healthy," "exotic," "tasty," and "seductive." As all these words carried a positive connotation except for "artificial," we ultimately made the decision to substitute "artificial" with "natural."

Spanning the space of properties

The Space of Properties lists the physical properties of the product that can have an effect in the elicited Kansei. Properties are usually called items, and the different values that an item takes in the study are called categories (for instance, one item could be colour, with categories red, blue and white).

Spanning the Space of Properties comprises the following steps:

1. Compiling an inventory of all conceivable tangible characteristics of a product and then opting for those that seem to exert the most significant influence on users' emotional perceptions.
2. Formulating the design matrix, which outlines the number of products designated for the experiment and determines the assigned value of each property for every individual product.
3. Picking out the products (or crafting product prototypes) in alignment with the specifications laid out in the design matrix.

Physical properties of the product that can be changed in production, are selected based on the experience and intuition of the team working on the project. The flowchart presented in Part 1 of this book can be used. A good idea is having people in the team giving different points to each one of the potential items, summing up all the points, and drawing a Pareto chart to select the items that had the most points in total.

In our study with fruit juices, five items were selected (***Table 14.1***)

Table 8.5: Factors and levels in the juices experiment.

Item	Categories
Straw	Yes
	No
Decoration	Yes
	No
Ice	Yes
	No
Container	Glass
	Goblet
Color	Yellow
	Orange

Once the items that will be used in the KE study are selected, the set of products that will be rated by participants must be prepared.

There are two different situations regarding the selection of products for the study:

1. When the products are created. In this case, products are prototypes or mock-ups (perhaps not fully functional). Products can also be created with renders, and presented as photos or videos. The advantage of this approach is that we are free to produce the prototypes that we want, and we can use sets of products that constitute a well-defined design matrix for further computations.

2. When existing products are used. In this approach, a set of existing products is collected trying to cover as many combinations of factors' levels as possible. The problem with this approach is that we can end with design matrices that are very far from balanced (orthogonality), and create confusion in the effects of the items once they are analyzed.

In our juice experiment, the prototypes were prepared according to the design matrix. The design matrix is a fractional factorial design (in particular, a 2^{5-1}). Table 8.6**Error! Reference source not found.** shows t he design matrix for the juice experiment.

Table 8.6: Design matrix for the fruit juice experiment.

	Straw	Decoration	Ice	Container	Color
1	No	No	No	Glass	Orange
2	Yes	No	No	Glass	Yellow
3	No	Yes	No	Glass	Yellow
4	Yes	Yes	No	Glass	Orange
5	No	No	Yes	Glass	Yellow
6	Yes	No	Yes	Glass	Orange
7	No	Yes	Yes	Glass	Orange
8	Yes	Yes	Yes	Glass	Yellow
9	No	No	No	Goblet	Yellow
10	Yes	No	No	Goblet	Orange
11	No	Yes	No	Goblet	Orange
12	Yes	Yes	No	Goblet	Yellow
13	No	No	Yes	Goblet	Orange
14	Yes	No	Yes	Goblet	Yellow
15	No	Yes	Yes	Goblet	Yellow
16	Yes	Yes	Yes	Goblet	Orange

Photographs representing each one of the prototypes were taken. The final prototypes used in the study are shown in Figure 8.3.

Figure 8.3: Prototypes used in the fruit juices experiment.

Data collection.

When dealing with data, the methodology of data collection holds immense significance, as always. The principle of "garbage in, garbage out" is applicable here: if the initial data gathered lack credibility, the conclusions drawn from our study will likely be inadequate or even incorrect, irrespective of the complexity of subsequent statistical analyses.

Prior to commencing the actual process of data collection, it becomes crucial to ensure that each participant comprehends the same meaning associated with every Kansei word. It is equally essential for all participants to recognize that their ratings hold equal value; there exist no right or wrong answers, and this isn't an examination.

In the context of the juice experiment, participants were provided with an introduction that elucidated the study's purpose and the procedure involved. A reference guide featuring all Kansei words and succinct explanations for each term was furnished prior to initiating the rating process. Participants were actively encouraged to peruse this guide and to inquire about any uncertainties they might encounter.

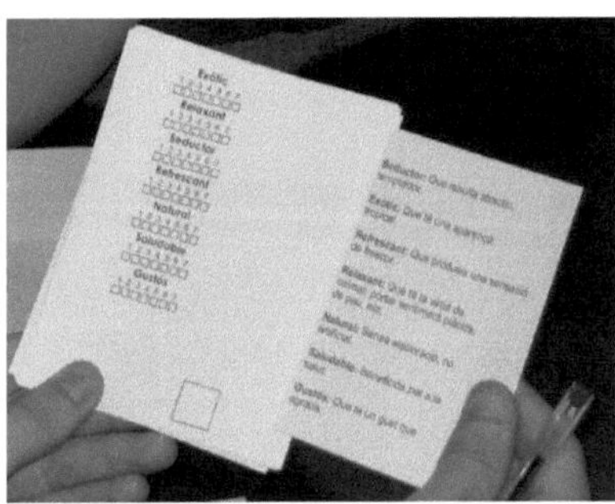

Figure 8.4: The legend with a definition of the Kansei words, and one of the forms for rating Kansei words.

The data collection has three dimensions: subjects, Kansei words and prototypes. Each participant is presented with a product. The participant rates this product on all the Kansei words. To avoid any bias, products and Kansei words are randomized as much as possible (see Figure 8.5).

Figure 8.5: Groups of 5 or 6 people rating the juices shown on the screen (Source: Pexels.com, Photographer: Christina Morillo).

Synthesis

After the process of data collection concludes, the subsequent step involves data analysis. The synthesis phase stands as the central component of Kansei Engineering, as it is here that the connection between product emotions and physical attributes is established. In situations involving quantitative data, statistical methodologies or automated learning techniques can be employed. The synthesis phase is connected to the next step, validation and model building: these two last steps should produce graphs that are easy to interpret and that can lead to action.

Let's first look at the data from the juices experiment using a simple tool: a radar chart. We've computed the average of all the ratings done by all persons to each one of the Kansei words and juices. Figure 8.6, left, shows the radar chart for the Kansei words Seductive and Exotic.

The numbers 1 to 16 correspond to each one of the 16 juices. The fact that the shape of the lines for both words (seductive and exotic) is very similar implies that these two Kansei words are perceived in a similar way (when a juice is perceived as being seductive, it's also perceived as exotic, and viceversa).

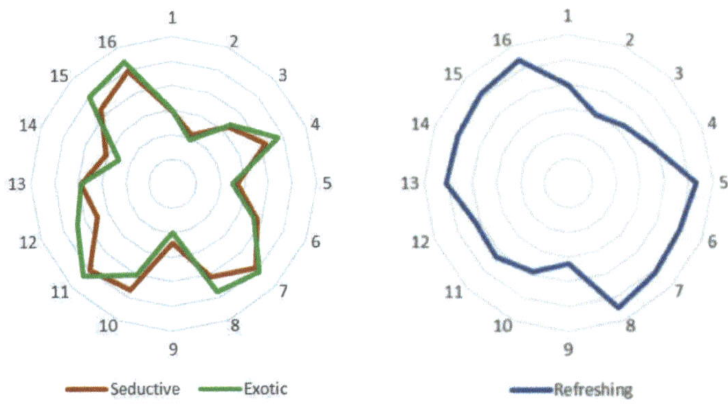

Figure 8.6: Radar charts for Kansei words Seductive, Exotic (left) and Refreshing (right).

Let's now look at the same graph for the Kansei word Refreshing (Figure 8.6, right). The values for Refreshing are quite low for juices 1, 2, 3, 4, 9, 10, 11 and 12. On the contrary, they are quite high for juices 5, 6, 7, 8, 13, 14, 15 and 16. If you recall Table 8.5 you will realize that juices 1, 2, 3, 4, 9, 10, 11 and 12 had no ice, whereas juices 5, 6, 7, 8, 13, 14, 15 and 16 had ice. Therefore, juices with ice have a high rating for the Kansei word Refreshing, and juices with no ice have a low rating for that Kansei word. This fact makes us think that having ice or not is a property that affects the perception of Refreshing.

This "manual" investigation can be automatized with statistical tools that facilitate analysis and further interpretation of results. Two tools frequently utilized are Factor Analysis and Quantification Theory Type

I (QT1). As these two tools are very common in Kansei Engineering studies, a section further in this text ("Some technical details on the analytical methods") will give a very brief overview of them.

Now, let's use them to discover insights from our data. Figure 8.7 shows the variables factor map from a Factor Analysis. This can be considered a representation of the semantic space. This kind of graphs place Kansei words that are perceived as similar close together, and those ones that are independent one from the other at about 90 degrees. For instance, words seductive and exotic are very close to each other, meaning that juices perceived as seductive are also perceived as exotic (this is something we already sensed with the radar chart for these two words).

Healthy, natural and relaxing are also words close together in the semantic map, meaning that juices perceived as natural are also perceived as healthy and relaxing.

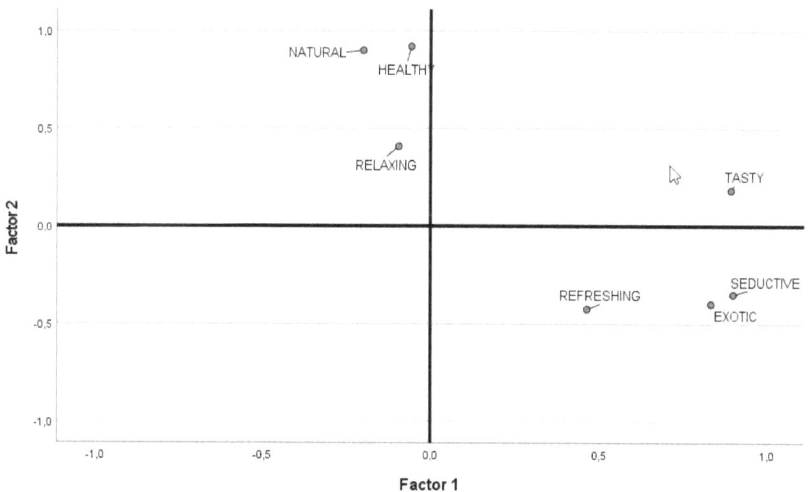

Figure 8.7: The variables factor map from a factor analysis with data from the juice experiment.

One of the most important outputs from many KE studies is discovering which properties affect each one of the Kansei words, and in which way (we are then linking the semantic space with the space of properties).

Quantification Theory Type I (QT1), a kind of regression analysis, is a useful tool for achieving this. The output from QT1 can also be converted into a graph, easy to interpret by designers and technicians. For instance, the QT1 results for the Kansei word Refreshing can be seen in Figure 8.8.

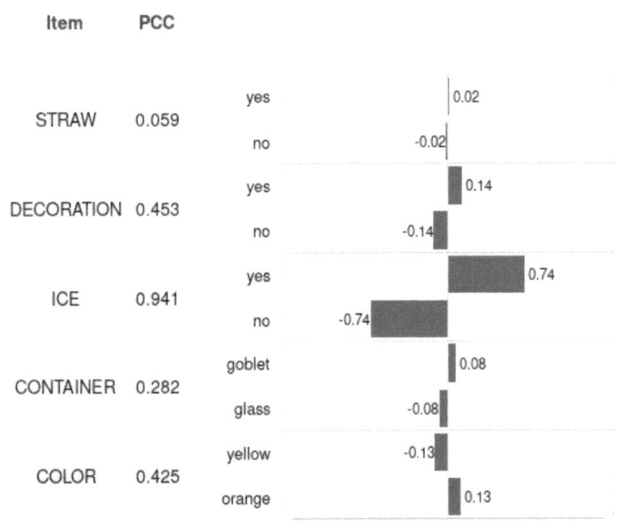

Figure 8.8: QT1 output for the Kansei Word Refreshing.

How can we interpret this graph? As we can see, we have each one of the properties studied on the left. The partial correlation coefficient (PCC) for each property measures the degree of association between the response (Refreshing) and each of the properties, taking into account that the effect of the other properties has been removed. In practical terms, when the PCC is high - say, over 0,7 – we can assure that the property has a real effect on the Kansei word. Another very good option is using a p-value for each property (when the p-value is low – say, less than 0,05 – the property is significantly affecting the response).

In our case, only the property Ice is affecting the perception of Refreshing. The numbers next to the bars in each of the categories of the property (Ice: yes 0,74; Ice: no -0,74) are called category scores. They show the differences in the rating referred to the average level of the response. In this case, having ice in the juice will increase the rating of Refreshing 0,74 points (on average) with respect to the mean; no having ice in the juice will decrease the rating of Refreshing 0,74 on average. Therefore, if we want our juice to be perceived as refreshing, having ice is the way to go.

After applying QT1 to each one of the Kansei words, we can summarize the results in a table (Table 8.7). This table shows which properties affect each one of the Kansei words (note that, for some Kansei words, we haven't found any property that affects).

Table 8.7: Summary of properties affecting each one of the Kansei words.

Kansei word	Properties affecting
Refreshing	Ice
Healthy	-
Exotic	Decoration Color
Seductive	Color Decoration
Natural	-
Relaxing	-
Tasty	Color

Test of validity and model building

This last step of the process includes confirming the conclusions of the synthesis phase (some confirmatory experiments could even be done), and presenting results in an understandable way that leads to action.

The presentation of results can be done with graphs that are easy to understand for everybody with a bit of effort, and that directly translate numerical data into useful information.

Descriptive graphs such as the ones presented in the last section (radar charts, profile charts) can be easily understood by everybody and give a first idea of the conclusions. A factor analysis of the Kansei words (as presented in the fruit juices experiment) can give a clear overview of the semantic space. More or less sophisticated statistical or machine learning tools can be used to establish a model that links the semantic space with the space of properties, one of the unique characteristics of KE studies. In particular, QT1 (as used in the fruit juices experiment), can be useful to present an easy to interpret graph that gives hints on how to set the technical properties of the product to elicit the desired emotional response.

Some technical details on the analytical methods

From the whole toolbox of statistical methods available and used in Kansei Engineering studies, there are two that stand out: factor analysis and regression analysis (in the form of Quantification Theory Type I, QT1). You need software to conduct these methods (refer to the next section on Software Solutions). But having a very basic understanding on how these methods work will help you in interpreting its results.

Factor analysis:

The idea of factor analysis is evaluating the correlation structure of a set of variables, reducing the whole set of variables into only a few dimensions (factors). Each one of these factors represent a latent variable that condenses the information of some of the initial variables. At the end, we only work with a few of these factors (the ones that capture the highest amount of information from the original variables). This is why we say factor analysis is a dimension reduction technique.

The algorithm for conducting a factor analysis implies computing the eigenvectors of the correlation matrix (you need software for doing this). But, at the end, the results are very often easy to interpret and very illuminating. Sometimes the latent variable expressed by a factor can be clearly understood based on the original variables most correlated with it, so we can "give a name" to each one of the factors.

A very nice graphical output from factor analysis is the loading plot for the first two factors (please look again at Figure 8.6, from the juice fruit experiment). This represents the semantic space in a way that can be easily interpreted.

Quantification Theory Type I (QT1):

QT1 is very commonly used in classical KE studies for linking the semantic space and the space of properties. The initial idea of QT1 comes from a seminal paper by Chikio Hayashi (Hayashi, 1952).

QT1 is basically a multiple linear regression analysis. In a multiple linear regression analysis we have one response (the dependent variable, the Y), and several regressors (independent variables, Xs). The aim is, first, discovering which independent variables Xs (from our list of candidates) affect the response Y. Secondly, we want to know how the variables Xs affect the response Y. At the end, we'll have an equation of the type $Y = \beta_0 + \beta_1 X_1 + \beta_2 X_2 + \cdots + \beta_k X_k$. The method for computing the coefficients of the equation is called ordinary least squares method.

QT1, however, has some particularities, it's not a common multiple linear regression. The most important issue is that the regressors are not continuous variables, but categorical variables (for instance, one regressor could be colour, with three different levels: red, blue and white). This implies that we need to use dummy variables to include these categorical variables in the model. Dummy variables make it possible to convert categorical variables into numbers (1, when the observation has that level, 0 otherwise). For instance, the categorical variable Colour, with levels red, blue and white could be converted into the dummy variables Colour_red (1 when it's red, 0 otherwise), and Colour_blue (1 when it's blue, 0 otherwise). We don't create Colour_white, as when Colour_red = 0 and Colour_blue = 0, this implies that the colour is white. In this way, we avoid having a situation of multicollinearity, something that would make impossible the computation using least squares.

The other peculiarity of QT1 is that, although using dummy variables, the model does not have a reference level for each property. All levels appear in the model, and the coefficients (called category scores) are referred to the general mean of the response. This makes interpretation of results much easier.

This was just a very brief overview of factor analysis and QT1. As we say in part 1 of this book, there are many other tools that can be used in KE studies. One very good way to deepen in the details of useful tools and learn about them is looking at the help facilities of statistical software (such as SPSS or Minitab). The help information of these statistical software packages is usually well produced and give more actionable advice on how to use the tools than more theoretical books.

Chapter 9: Software Solutions

Kansei Engineering Software (KESo)

In response to the lack of automated tools specifically available for Kansei Engineering evaluations, Linköping University has developed a universal software called Kansei Engineering Software (KESo) for data collection and basic evaluation. The goal was to create a tool that could automate some of the processes requiring special knowledge in statistics or behavioral sciences. (*Kansei Enginering Software- KESo*, 2017; Schütte et al., 2006);

KESo is an online tool that can create a data collection form for Kansei Engineering, collect data, and store them in an internal database. In the next step, KESo can evaluate the collected data using various mathematical evaluation methods such as Quantification Theory Type 1, Rough Sets Theory, and other linear methods. From there, models can be built to quantify the link between the subjective, affective user impression and concrete product properties fine adjustment.

The KESo software is based on the general Kansei Engineering model depicted in Part 1 Figure 4.1 and uses its logic and procedures. The software is available online at www.kanseiengineering.net. KESo is a powerful tool that enables researchers to conduct Kansei Engineering studies with ease, even without specialized knowledge in statistics or behavioral sciences.

KESo is often seen as a significant contribution to Kansei Engineering research, as it enables researchers and practitioners alike to conduct studies with ease, automate some of the processes requiring specialized knowledge, and provide multiple mathematical evaluation methods for data analysis. The availability of KESo online and the provision of a tutorial video make it accessible to a broad range of researchers and practitioners in the field of Kansei Engineering.

Kansei Engineering methodology usually contains a number of complex processes which require significant resources and interdisciplinary knowledge in areas such as engineering, psychology, mathematics, and statistics. However, the complexity of this process has hindered its application and integration into industrial processes, making it challenging for many companies to conduct such studies.

While larger companies in industries such as vehicle manufacturer have the necessary equipment, knowledge, and manpower to conduct regular affective evaluations of their products and develop their own tools, these methods are often considered competitive and not widely shared outside of these companies. This lack of access to resources puts smaller companies at a disadvantage in comparison to their competitors.

Despite recognizing the potential benefits of affective evaluation tools, many smaller companies simply cannot afford them. As a result, potential tools for affective assessment must be accessible, user-friendly, affordable, and not require specialized knowledge. These requirements are essential to enable small and medium-sized enterprises (SMEs) to conduct basic Kansei Engineering studies and improve their competitiveness in the market.

In response to these challenges, this chapter presents some of the most commonly used software solutions for conducting basic Kansei

Engineering studies. By providing SMEs with these tools, companies can gain valuable insights into customer preferences and improve the design of their products, ultimately leading to increased customer satisfaction and market success. Use examples of some of these Kansei Engineering Software solutions can be found in Parts 3 and 4.

Additional Software solutions

While the above mentioned KESo Kansei Engineering Software is the only expert software on Kansei Engineering in existence, several other commercial software packages are useful and are frequently used in the context of Kansei Engineering methodology. The list presented here is not exclusive but rather a selection of tested tools for Kansei Engineering cases. The list starts with the most affordable and accessible software moving towards tools used by specialists.

Spreadsheet software

Spreadsheet software allows users to organize, analyse, and manipulate data using rows and columns. It provides various features such as mathematical functions, charts, graphs, pivot tables, and macros to help users manage and interpret their data. Excel is widely used in industries such as finance, accounting, marketing, and engineering for tasks such as budgeting, forecasting, data analysis, and reporting.

In the context of Kansei Engineering it is often considered as the first choice when performing data compilation and modification of the data collected. Also minor visualizations or even evaluations can be done. In several cases users have programmed so called macros in order to facilitate the conditioning of the incoming data. In Parts 3 and 4 spreadsheet software is extensively used to analyse the example data and for teaching purposes. Despite the general nature of this software solution, it is quite versatile in the Kansei Engineering context.

SPSS

SPSS (Statistical Package for the Social Sciences) is a software program used for statistical analysis in social sciences, such as psychology, sociology, and political science. It allows users to input data and perform various statistical tasks such as t-tests, ANOVA, regression analysis, and factor analysis. SPSS also provides graphical representation of data, allowing users to visualize relationships between variables. SPSS is widely used in research and academic settings for data analysis, survey research, and statistical modelling. It is also used in business and government sectors for market research, customer satisfaction surveys, and policy analysis.

Since SPSS is a more specialized software focused on statistical data analysis, it yields more in depth insights to Kansei Engineering data. Although it requires a fair amount of knowledge in statistics and a bit of training to be used in a good manner, it must still be considered easily accessible to the occasional user.

In Kansei Engineering SPSS is often used for examining the data quality, (e.g. ensuring normally distributed data, detecting anomalies in the collected dataset, finding correlations between the collected data, etc.). Also various data visualizations can be carried out when e.g. performing Factor analysis.

MINITAB

Minitab is a statistical software package designed for data analysis and quality improvement. It provides a wide range of statistical tools, including basic descriptive statistics, hypothesis testing, regression analysis, and design of experiments. Minitab is widely used in industries such as manufacturing, healthcare, and finance to improve product quality, reduce costs, and increase efficiency. It also offers graphical and visualization tools, making it straightforward to present and interpret statistical data.

Minitab Software tool

While Minitab is known for its user-friendly interface and easy-to-learn features, it still requires a good amount of knowledge of statistical methods to get the best from it. From the point of view of a product developer using Kansei Engineering, this software package is a good option as it is commonly used in industry and therefore easy to integrate into an industrial user's workflow.

For Kansei Engineering it can help to analyse the incoming user data, perform quality checks and data visualization.

R- Software

R is a free of charge programming language and software environment for statistical computing and graphics. It provides a wide variety of statistical and graphical techniques, such as linear and nonlinear modelling, time-series analysis, clustering, and more. R is an open-source software, which means it is free to use and can be customized and extended by its user community. It has a vast collection of user-contributed packages that provide additional functionalities, making it a powerful tool for data analysis, machine learning, and visualization. R is widely used in academic research, data science, and industry for various purposes, such as statistical analysis, predictive modelling, and data visualization.

In comparison to the previous software packages, R-software requires a sound knowledge of statistics and programming. It requires a substantial amount of training to be used in a sufficient manner and hence, it is more suited to expert users. R-software offers a range of add-ons for data visualization that can be used in Kansei Engineering.

In the context of this book, the examples will not require R-software knowledge, but be advised that it was used in the analysis of several of the studies presented in Part 4.

Python

Python is a high-level, interpreted programming language that is widely used for general-purpose programming, data analysis, web development, and artificial intelligence (AI). It is known for its simplicity, readability, and flexibility, making it a popular choice for beginners and experienced programmers alike. Python has a large standard library that provides various functionalities for tasks such as file manipulation, networking, and web development. It also has a vast collection of third-party libraries and frameworks that make it easier to work with data, create web applications, and build machine learning models. Python is an open-source language, which means it is free to use and can be customized and extended by its user community. It runs on multiple platforms, including Windows, Mac OS, and Linux, and is supported by a large and active developer community.

Clearly Python is a most versatile software tool, but requires the highest amount of expertise in programming and statistics as well as specialized statistical methods used within Kansei Engineering. Python is used to program own software tools for Kansei Engineering purposes and early versions of the KESo Kansei Engineering Software were based on Python.

References for Part 2

Hayashi, C. (1952). On the prediction of phenomena from qualitative data and the quantification of qualitative data from the mathematico-statistical point of view. *Annals of the Institute of Statistical Mathematics,* vol. 3, no. 2, pp. 69.

Jindo, T., Hirasago, K., & Nagamachi, M. (1995). Development of a design support system for office chairs using 3-D graphics. *International Journal of Industrial Ergonomics, 15*(1), 49-62.

Kansei Enginering Software- KESo. (2017). Linköping University. www.kanseiengineering.net

Lokman, A. M., Awang, A. A., Omar, A. R., & Abdullah, N. A. S. (2016). The integration of quality function deployment and Kansei engineering: an overview of application. *AIP Conference Proceedings* (Vol. 1705, No. 1). AIP Publishing.

Marco-Almagro, L. (2011). *Statistical Methods in Kansei Engineering Studies.* UPC Barcelona Tech.

Mazur, G. (2000). QFD 2000: Integrating QFD and Other Quality Methods to Improve the New Product Development Process. *12th Symposium on QFD/6th International Symposium on QFD2000. Proceedings of 12th Symposium on QFD/6th International Symposium on QFD2000* (pp. 305-317).

Picard, R.W. (2000). *Affective computing.* The MIT Press.

Schütte, S., Eklund, J., Axelsson, J. & Nagamachi, M. (2004). Concepts, methods and tools in Kansei engineering. *Theoretical Issues in Ergonomics Science,* vol. 5, no. 3, pp. 214.

Schütte, S. (2005). *Engineering Emotional Values in Product Design: Kansei Engineering in Development*, Linköping University.

Schütte, S., Alikalfa, E., Schütte, R., & Eklund, J. (2006). Developing Software Tools for Kansei Engineering Processes: Kansei Engineering Software (KESo) and a Design Support System Based on Genetic Algorithm. *QMOD 2006*.

Schütte, S., & Schütte, R. (2009). *Kansei Engineering Software (KESo)* (p. Data collection and evaluation software for Kansei). Linköping Institute of Technology. www.kanseiengineering.net

PART 3

Kansei Engineering Product Stories

The authors of this textbook have a combined long term experience using Kansei Engineering in practice in academic and industrial contexts. In this part several of their stories are told to demonstrate the reach of the method and the versatility of its use. The stories are retold in different styles reflecting the different types of application and different stakeholder groups. Further details of each of the stories can be found in additional publications listed in the bibliography.

Chapter 10: Chocolate Snacks - An example from food industry

By Simon Schütte and Lluis Marco Almagro

This case study was carried out by researchers at Linköping University, Sweden and at UPC Barcelona, Spain in collaboration with leading confectionary manufacturer in Scandinavia. The study results or parts of them were published at several occasions and in different media. The main study was carried out as a research project funded by the Swedish governmental research entity VINNOVA reported in Simon Schütte & Marco-Almagro (2013). Also several master's theses carried out by Grzechnik & Priithiviraj (2010), Prithivirat & Grzechnik (2011) and Friberg (2010) used parts of the data and delivered further insights. A journal paper by Schütte (2013) used the findings of the study to further examine how chocolate packaging can affect customer expectations on chocolate snacks and derive guidelines on how the interior must be designed in order to fulfil those expectations. In 2022, Schütte and Marco-Almagro published a comprehensive model of the adapted general Kansei Model joined with the Food Kansei Model (Ueda et al., 2008)

The company involved is one of the leading confectionary manufacturers in northern Europe. Amongst other products they produce two sister products containing several layers of wafer covered with chocolate.

Figure 10.1: Chocolate confectionary sample picture (Source: Pexels.com, Photographer: Terrance Barksdale).

Both products are very similar, but one has a thinner chocolate cover and is selling better to a wider market. One product name suggests a closer connection to on-the-go consumption while the other is more perceived as a leisure product. The first one seems to sell better to male consumers while the latter targets a wider audience. With this in mind the development team was eager to differentiate these products further giving the first product its own affective identity towards sports as well as opening it to a wider audience (among other things eradicating the male preselection).

Aim of the study

The objective of this study was to identify a toffee-filling possessing an emotional characteristic aligning with the desired brand perception of Sportlunch as both athletic and invigorating. To elaborate further, the goal was to discover an appropriate blend of components that is not only favored by consumers but also aspired to. Additionally, a secondary objective involves pinpointing a combination that is likely to be effective in multiple European countries, extending beyond just Sweden.

Study structure

The study was carried out in accordance with the modified Kansei Engineering food model as presented in ***Figure 10.2***. This model is a combination of the Kansei Food model (e.g. Ueda et al., 2008) and the general Kansei Model as presented in Part 1 of this book. The hybrid model is described in Schütte & Marco-Almagro (2022).

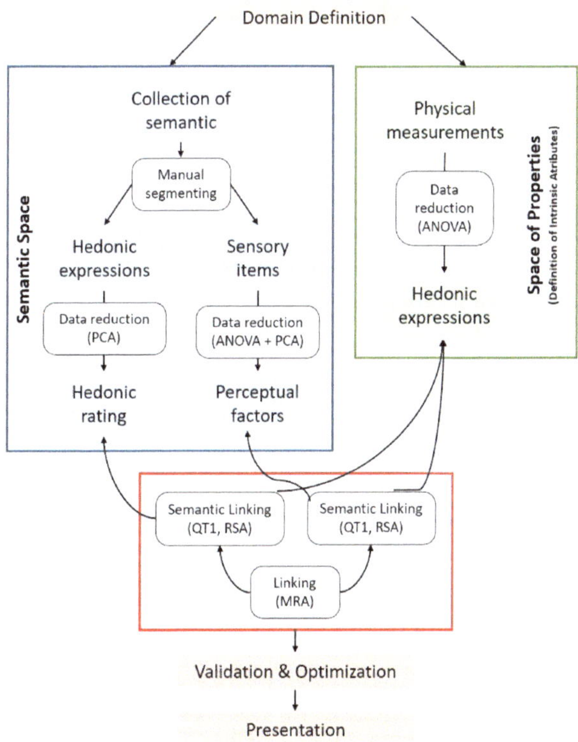

Figure 10.2: Modified Kansei Engineering model for food industry **(Schütte & Marco-Almagro, 2022).**

In the initial stage, a domain was selected, from which both the Semantic Space and the Space of Application were defined. For the establishment of the Semantic Space, emotional words were gathered. These words were drawn from earlier research studies (Grzechnik & Priithiviraj, 2010; Prithivirat & Grzechnik, 2011; Schütte, 2013b), advertisements for competing products showcased during sports events, and input from the company's own development team. To enhance this dataset, a preliminary study was conducted, involving 25 consumers who were asked to verbally describe their impressions while consuming Sportlunch. These verbal inputs were then manually reviewed and categorized into expressions related to emotional experience and sensory attributes. By employing the affinity diagram technique, the list of semantic terms was condensed, and phrases with overlapping meanings were removed.

Concurrently, the product development division within the company identified key ingredients and potential combinations that had a significant emotional impact. Through collaboration with Linköping University in Sweden, an experimental plan was formulated, and the chosen prototypes were manufactured.

Kansei Engineering Software (KESo) was used for creating a data collection homepage. Up front the participants were screened by providing information such as: Age, Gender and frequency of chocolate consumption as well as their test group code (nationality). In the actual data collection the consumers were provided with an assortment box containing samples of individually numbered chocolates containing different samples. Despite the different taste the samples looked identical. In the following test phase, the participants were asked to proceed as follows: "Put the sample in their mouth and crack it open by biting on it. Leave the filling on the tongue for a short while before evaluating their personal impressions on special VAS scales in a position which “feels” right”.

Figure 10.3, left shows an example of the data collection site in KESo Software.

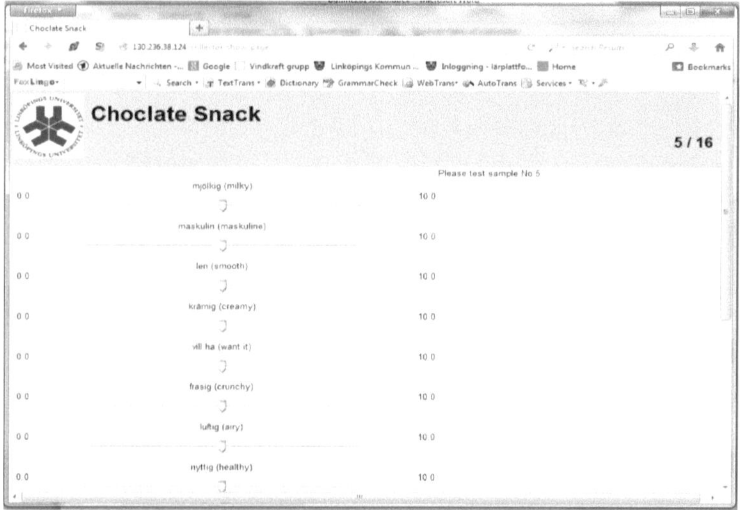

Figure 10.3: Product evaluation using KESo software.

The data gathering process occurred in two distinct phases. Initially, the survey was completed by Swedish individuals who were regular to heavy consumers of chocolate. Subsequently, the same survey was responded to by Spanish chocolate consumers during the second phase. See Figure 10.4.

Figure 10.4: Data collection using KESo (Source: Pexels.com, Photographer: Christina Morillo).

KESo software provided instant results by applying QT1 analysis (Komazawa & Hayashi, 1976) and Rough Set Analysis (Nishino et al., 2001). Subsequently the data was checked for data quality and consistency. Also the data was treated with new not yet proven analysis methods such as Ordinal Logistic Regression (OLR) (Marco-Almagro, 2011).

Study results

As a result of the semantic data collection a total number of 151 affective verbal expressions were collected. Of those, 55 expressions were identified as hedonic expressions, 32 were sensory items (20 described the texture and 12 the taste). The rest were either variants of expressions (sweet, sugary, sugar, etc) or deemed irrelevant. After the affinity analysis (Bergman & Klefsjö, 2002) the words in Table 10.1 remained and were used in the following data collection.

Table 10.1: Kansei Words defining the Semantic Space.

<table>
<tr><th colspan="2">Kansei word type</th><th>Kansei words</th></tr>
<tr><td colspan="2" rowspan="5">Hedonic expressions</td><td>Want to have</td></tr>
<tr><td>Healthy</td></tr>
<tr><td>Energy</td></tr>
<tr><td>Masculine</td></tr>
<tr><td>Like it</td></tr>
<tr><td rowspan="7">Sensory items</td><td rowspan="4">Texture related</td><td>Smooth</td></tr>
<tr><td>Creamy</td></tr>
<tr><td>Crunchy</td></tr>
<tr><td>Airy</td></tr>
<tr><td rowspan="3">Taste related</td><td>Milky</td></tr>
<tr><td>Nut-like</td></tr>
<tr><td>refreshing</td></tr>
</table>

In parallel, the Space of Properties was defined. Several ingredients deemed by the companies taste developers as possessing a high affective impact were chosen and combined in a dummy coding table as shown in Table 10.2 below. The ingredients chosen were caffeine, three different flavours, yoghurt and chilly spice to produce a hot feeling. Table 10.2 shows the 12 combinations which were tested in the main study.

Table 10.6: Space of Properties as defined in a Dummy Coding Table.

Sample No.	Caffeine	Flavour	Yoghurt	Spice
1	Yes	Salt	No	Chilli
2	Yes	Salt	No	None
3	No	Salt	Yes	Chilli
4	No	Salt	Yes	None
5	No	Nut	No	Chilli
6	No	Nut	No	None
7	No	Nut	Yes	Chilli
8	No	Nut	Yes	None
9	Yes	Fruit	No	Chilli
10	Yes	Fruit	No	None
11	No	Fruit	Yes	Chilli
12	No	Fruit	Yes	None

Data analysis results

Data quality

The data collected in KESo software were checked in accordance to their quality. A K-S test was performed to check whether the data was normally distributed. That was the case for all Kansei words used. Also, it was checked if the gender, the age and the nationality (Swedish, Spanish) made a difference in how the test samples were perceived. As for the gender, significant differences were found only for the word "smooth in sample 1". Otherwise no significant differences were found for gender. No significant differences were found for age. Worth mentioning here is that the age range was quite narrow. For the nationality, the Levene's Test showed 3 significances regarding the variances. However, no significant differences were found for the means of the nationality rankings in the t-test. Based on these results, the data was deemed satisfactorily homogeneous and pooled for further evaluation.

Affective fingerprints

Each taste sample was rated on a 100 degree scale (VAS scale) and thereby obtained its own individual affective profile. ***Figure 10.5*** displays the average ratings plotted into a radar chart. It can be seen that the emotional profiles typically show good distribution and wide variability. This is a sure sign that the taste of the sample is well distinguishable. Hence it can be assumed that the ingredients chosen for spanning the Space of Properties were a sensible choice.

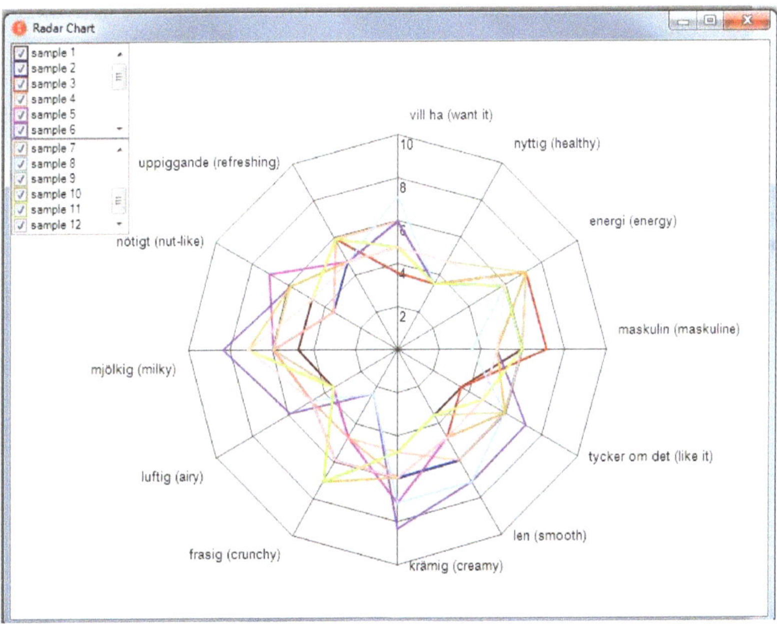

Figure 10.5: Emotional Profiles for each sample; all participants' average.

Analysis of the Semantic Space

A principal component analysis (PCA) can give deeper insight into the respondent's way of thinking. The data gathered were examined separately on sensory items and hedonic expressions and Swedish and Spanish respondents. The results from the different countries were fairly similar. Hence the data were pooled and a combined PCA conducted strengthening the validity. ***Figure 10.6*** below shows the result. The closer the words are together, the more alike are they perceived by the participants.

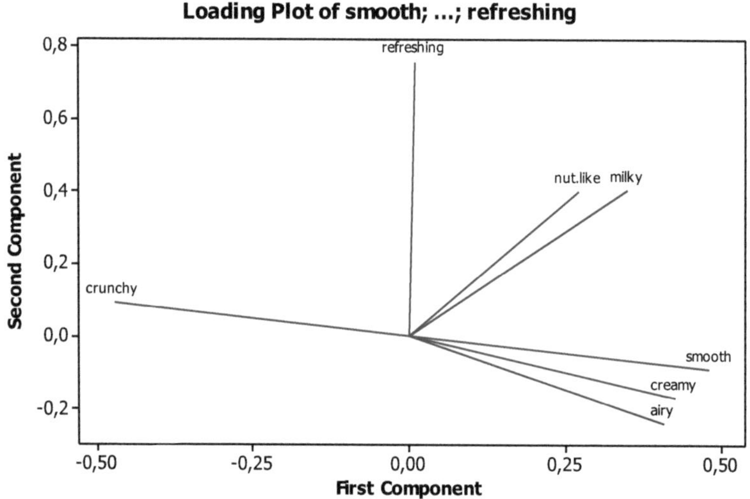

Figure 10.6: The Semantic Space for sensory items- Mathematically defined.

In this study creamy, smooth and airy fall very close in the Semantic Space and are positioned opposite crunchy. A product can either be smooth and creamy or crunchy but not both at the same time. Nutlike/milky and refreshing turn out being independent of each other and the other dimensions.

The fact that the deployment of this semantic space of sensory items is very similar in both countries makes sense. Participants in the study seriously rated all sensory items and arrived at similar conclusions in both countries, as these sensory items are intrinsic attributes of the chocolates and easier to "objectivize".

The PCA for the hedonic expressions paints a more dissimilar picture. Most hedonic expressions fall far away one another. The exception is for "want to have" and "like it" for the Spanish results, where both are quite close. See Figure 10.7.

The similar results for the sensory items can be explained with similarities in previous experiences with textures and tastes. The hedonistic results however vary due to the fact that Spanish respondents are not used to the Swedish brand flavour preferences. This does not, however, mean that the product cannot be sold in different European countries, but that they will display different hedonic values to the consumers.

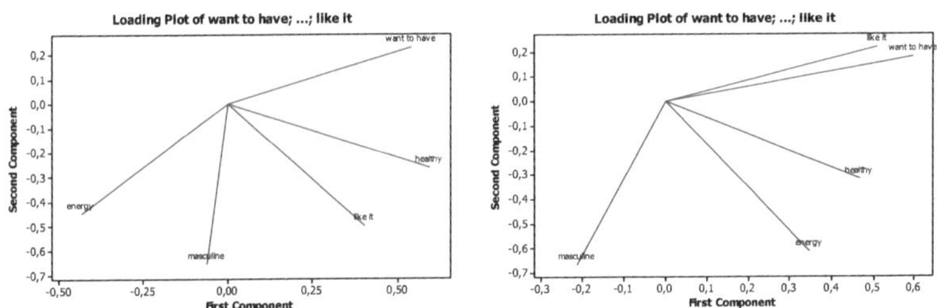

Figure 10.7: The Semantic Space for hedonic expressions; left Swedish and right Spanish results.

A linear regression analysis has been done using each one of the hedonic expressions (want to have, like it, healthy, energy and masculine) as responses, and the sensory items (smooth, creamy, crunchy, airy, milky, nut-like and refreshing) as possible regressors.

The selection of the best model for each response has been done using a best subsets procedure (doing all possible regressions and choosing the best one based on squared R^2 and Mallow's Cp indicator).

The following Tables 10.3 summarize the results for both Swedish and Spanish data collections. The number "% response explanation" is the value of adjusted R^2 in the linear regression analysis. It represents the percentage of variability of the response that can be explained by the given variables.

Table 10.7: Relationship between hedonic expressions and sensory items based on regression analysis.

Sweden

	Smooth	Creamy	Crunchy	Airy	Milky	Nut-like	Refreshing	% response explanation
Want it	+	−						54
Like it				−		−		39
Healthy				−		−		54
Energy	−	+						30
Masculine			+		+		+	69

Spain

	Smooth	Creamy	Crunchy	Airy	Milky	Nut-like	Refreshing	% response explanation
Want it	+				+			61
Like it			−		+			76
Healthy		−		+		+		57
Energy			−		−	+		29
Masculine	−	−				+	−	77

The following are two examples to aid interpretation of these tables:

- Response Healthy, for Sweden: chocolates with low airy and low nut-like ratings are perceived as more healthy. However, these two sensory items explain only 54% of the perception of being healthy (therefore there are other issues that affect the perception of healthiness).

- Response Like it, for Spain: chocolates with low crunchy and high milky ratings have higher rating on the expression “like it”. And more than 75% of this preference choice can be explained by just these two sensory items (crunchy and milky).

Linking the Kansei words to the ingredients

There are many ways to link the Kansei words (sensory and hedonic items) constituting the Semantic Space to the selection of ingredients constituting the Space of Properties. Many of them are presented in Part 1 of this book. Also in this study several of those methods were used and their results compared to each other. Those were Ordinal Logistic Modelling, Rough Set Analysis and Quantification Method Type 1. The results can be found in detail in Schütte & Almagro (2013). Quantification Theory Type 1 (QT1) is the most commonly applied method and for the purpose of this book it will be the only one described here.

The evaluation using QT1 itself yielded a great amount of data. A typical QT1 output is displayed in Figure 10.8. It compares the displays for the hedonic terms “want it” and “like it”. The diagrams in each figure can be interpreted in the way that the combination of caffeine, nut taste, Yoghurt and no chilli spice will maximize the rating for “want it”. The picture is similar for “like it” where the combinations of no caffeine, nut or fruit taste, no yoghurt and no chilly spice produce the highest ratings. This means that a product which people like AND want needs to have nut flavour, and no chilli spice. It also shows that liking and wanting are different concepts when it comes to food products. KESo software can provide detailed data of this kind.

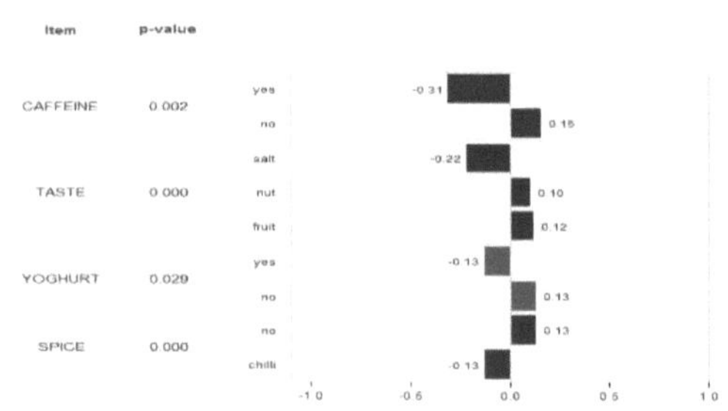

Figure 10.8a: Example QT1 output from all pooled data for "want it".

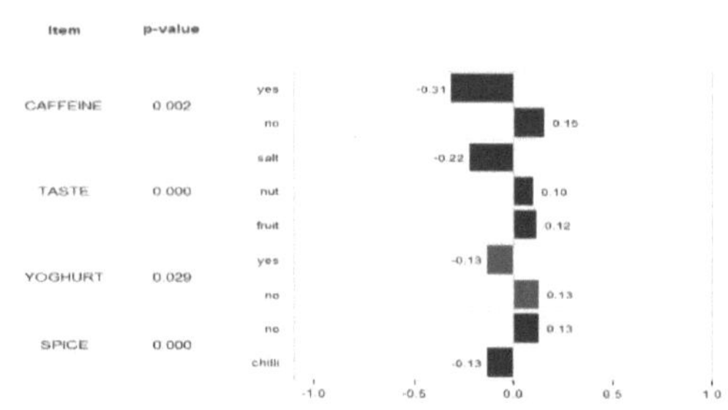

Figure 10.8b: Example QT1 output from all pooled data for "like it".

For obvious reasons not all Kansei words in the Semantic Space can be shown as in the figure above. Table 10.4 displays all the QT1 results in a condensed way. It can be seen that in some points there are differences in the ratings of the Spanish and Swedish groups. However, it can also be seen that despite those differences it is possible to find a common pattern for both groups. This implies that a chocolate product designed for either market probably will be perceived similarly on the other market as well.

Table 10.8: Assembly of all QT1 results. "+"/"-" means positive impact of the respective category on the average rating.

		Kansei word	Dataset	Caffeine		Taste			Yoghurt		Spice	
				Yes	No	Salt	Nut	Fruit	Yes	No	N	chilli
Hedonic expressions (Kansei Words)		Want it	Sweden	+	-	-	+	-	+	-		
			Spain			-	+	-	+	-	+	-
			All	+	-	-	+	-	+	-	+	-
		Like it	Sweden	-	+	+	-	+			-	+
			Spain	-	+	-	+	+	-	+	+	-
			All	-	+	-	+	+	-	+	+	-
		Healthy	Sweden	+	-	+	-	-	+	-		
			Spain	-	+		+	-			+	-
			All	+	-	+	+	-	+	-	+	-
		Energy	Sweden	-	+	+	-	+	-	+	-	+
			Spain	+	-	-	+	-	+	-	-	+
			All	-	+	-	+	-	-	+	-	+
		Masc-uline	Sweden	+	-	0	-	+	+	-	-	+
			Spain	-	+	+	-	-	-	+	-	+
			All			+	-	0	-	+	-	+
Sensory items	Textures	smooth	Sweden	+	-	-	+	-	+	-	+	-
			Spain	-	+	-	+	+	-	+	+	-
			All	+	-	-	+	-	+	-	+	-
		Creamy	Sweden			+	+	-	-	+		
			Spain			-	+	0	-	+	+	-
			All			0	+	+	-	+	+	-
		Crunchy	Sweden	+	-	-	-	+	-	+	-	+
			Spain	+	-	+	-	+	+	-	-	+
			All	+	-	+	-	+	+	-	-	+
		Airy	Sweden	+	-	-	+	-	+	-	+	-
			Spain	-	+						+	-
			All			-	+	-	+	-	+	-
	Taste	Milky	Sweden			-	+	-	+	-	+	-
			Spain	-	+				-	+	+	-
			All	-	+	-	+	-			+	-
		Nut-like	Sweden	-	+	-	+	0	-	+	+	-
			Spain	-	+	-	+	+	-	+	-	+
			All	-	+	-	+	+	-	+		
		Refreshing	Sweden			+	-	-			-	+
			Spain			-	-	+	+	-		
			All			+	-	-	+	-	-	+

Discussion of the Chocolate snack case study

When collecting data, a number of demographic questions were also asked. This was done to make it possible to stratify the data afterwards. Except for the citizenship, it was difficult to find subgroups regarding age or chocolate eating habits. The reason for this might be twofold. Partly, the low number of participants did not allow clear results. Partly, the group of participants was deliberately chosen as homogenous as possible. This meant that e.g. the age range was very narrow not allowing any valid conclusions.

Initially, it was planned to establish the links between the Semantic Space and the Space of Applications with three evaluation methods, namely, QT1, OLR and mixed model OLR (mOLR). The latter is according to Marco-Almagro (2011) the most sensitive method. Regrettably, the mOLR required a great amount of preparation. Considering the short amount of time to collect and evaluate three different products for this VINNOVA project, it was decided to only use QT1 and OLR. Noteworthy differences between those methods were found, so conclusions could be drawn. mOLR would have probably strengthened those conclusions.

Conclusions of the case

Considering the combined QT1 and OLR results, a number of similarities can be found. In order to realize a good product all semantic terms were considered; in particular the terms "want it", "like it" to ensure that consumers will repeatedly buy the product. Also the words "masculine", "healthy" and "refreshing" were chosen by the company's development team to be prominent for this product group. The rest of the words were chosen in a way that they had a positive influence on the product.

The best compromise with those prerequisites was:

Sweden: Caffeine: yes, Taste: fruit (or possibly salt), Yoghurt: Yes and Spice: No (only to emphasise masculine can chilli be added)

Spain: Caffeine: No, Taste: nut (or possibly fruit), Yoghurt: Yes and Spice: No (except for masculine)

Considering those selections, it is not impossible to find a combination which works for both countries. It can be speculated that such products might work for more than just the Swedish and Spanish markets.

Chapter 11: Apartments

By Simon Schütte

This study was carried out in cooperation with a multinational construction company. It has significantly contributed to the development of the country's essential facilities such as roadways, energy generation facilities, and residential buildings. Presently, the organization is recognized as a prominent participant in numerous nations. The company has contributed to this study with expertise of the apartment market as well als providing apartments for affective evaluations.

Constructing houses demands a significant amount of available funds during the building phase. To mitigate the investment's return risk, The constructor, like many other construction firms, enters agreements with most (private) homebuyers before commencing construction. As the outcome remains uncertain, customers often feel uncertain about the financial, emotional, and societal aspects. In simpler terms, prospective homeowners are unsure if they'll enjoy their new residence and if it's worth the cost. Thus, construction commpanies ai to address these concerns beforehand by enhancing how they present new properties, offering potential owners a sense of what it's like to inhabit the upcoming dwelling before it's physically established.

Various approaches were taken to examine the question, aiming to provide diverse viewpoints on the issue. A range of tools from various fields and disciplines were utilized. Kansei Engineering was selected as an appropriate assessment method for quantifying the impressions held by potential customers (compare: Bergqvist & Domeij, 2001).

Kansei Engineering technique was utilized to compare how three distinct ways of showcasing apartments (Virtual Reality, computer desktop presentations, and screen projections) along with the actual apartment and construction drawings, influenced the user experience. Refer to Figure 11.1 for visualization. The constructor's objective was to enhance the apartment presentation by incorporating Virtual Reality methods that most accurately captured the emotional response associated with the new apartment.

Figure 11.1: Example of a 3D rendering of an apartment.

Structure and layout of the experiment

The experimental structure adhered primarily to the principles outlined in the initial section of this book. Preliminarily, the constructor had already completed crucial aspects of Kansei Engineering. The domain, such as the target audience, product category, and market niche, had been chosen and defined beforehand. Furthermore, company staff had clearly articulated the specific product parameters of interest. Consequently, significant time was conserved by utilizing this data as input for an adapted Kansei Engineering investigation. Nonetheless, three significant aspects of

Kansei Engineering (compare part 1 in this book) still needed to be executed:

1. Spanning the Semantic Space
2. Synthesis
3. Model building

Spanning the Semantic Space

A total of 520 adjectives pertaining to both apartments and the presentation of apartments were gathered. Initially, this count was reduced by excluding adjectives that described the surroundings and circumstances. Subsequently, the remaining words were meticulously grouped into 21 clusters, and out of these clusters, 21 words that best represented the emotional responses (here visualized as Kansei words in Table 11.2) were selected.

Synthesis

Since the various methods of showcasing the products had already been selected, there was no need to deliberate on the Scope of Applications. During the synthesis phase, a combined assessment of the apartments and varied VR presentations was conducted involving 58 participants. To streamline the evaluation process for each respondent, a decision was made that each participant would assess only one VR presentation in addition to the construction plans and the actual apartment. The layout of the experiment is depicted in Figure 11.2 below.

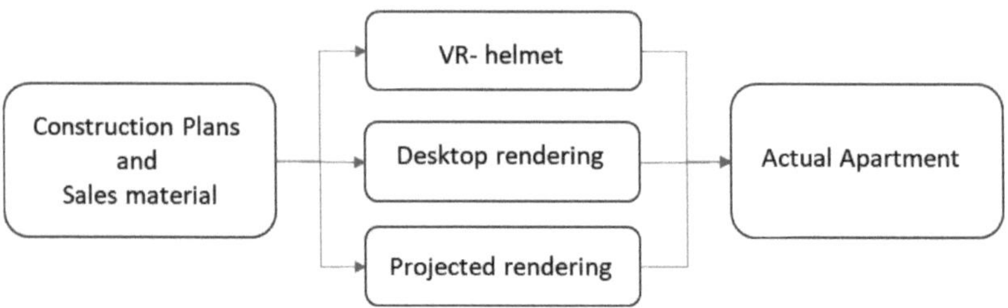

Figure 11.2: Study plan apartments (adapted from Bergqvist and Domeij, 2001).

During the evaluation phase, an initial factor analysis is conducted to reveal the perspectives through which prospective customers perceive newly constructed apartments. Subsequently, the Kansei words are associated with the distinct presentation methods. Unlike previous instances, the linkage between Kansei words and VR models isn't established using QT1 or a comparable approach. Instead, statistical tests are employed to compare the VR models with the actual apartment. This analysis indicates the areas in which the VR models align with or deviate from reality.

Model building

Because a QT1 analysis was not executed, it's not feasible to create a corresponding mathematical model as illustrated previously. Nevertheless, the outcomes reveal the perception of the VR models relative to the actual apartment. Based on this, specific observations can be formulated, serving as the derived model in this particular scenario.

Results

From the raw data a factor analysis with Principial Component Analysis was performed. The resulting factors were treated with Varimax rotation making the resulting factors independent of each other. This can be interpreted as if the prospective buyers of the apartments value a number of aspects their new apartment must rank high on. Those aspects (factors) are presented in Table 11.1 below.

Table 11.1: Resulting factor analysis (compare Bergqvist and Domeij, 2001).

Factor 1 is the **concrete factor**: It includes Kansei words like practical, well-planned, large, spacey, modern, etc.
Factor 2 is called **exclusivity factor:** Kansei words like original, fancy, elegant and luxurious.
Factor 3 is the **ambiente factor:** Kansei words like personal, comfortable (snug cosy), warm, colourful and tasteful are collected here.
Factor 4 expresses how the respondents **enjoy** the apartment: Dominating Kansei words are secure, enjoyable, inspiring.
Factor 5 describes **lighting and harmony:** Besides the words light and harmonic, even words like new and fresh are included.

In a second evaluation of the raw data the correlation of the ratings of the real existing apartment and the three ways of virtually simulating the apartment was calculated. The complete resulting data can be found in Bergqvist and Domeij, 2001. Table 11.2 below depicts an simplified aggregation of those results. A "+" means that the respective means of simulation yields a higher rating than the real apartment, a "-" a lower. A "0" indicates not significance.

Table 11.2: Connecting Kansei words to the way of presentation A negative value means that the presentation model is experienced as more negative than the real apartment regarding the Kansei word.

	VR- presentation		
	Googles	Desktop	Projection
spacious	+	-	-
well- planned	-	-	0
practical	0	0	0
big	0	-	0
modern	-	-	0
original	-	-	0
luxurious	-	-	0
cool	0	0	0
elegant	-	-	-
cozy	-	-	-
personal	-	0	-
colourful	-	0	0
warm	0	-	0
prestige	-	-	-
secure	0	-	0
enjoyable	-	-	-
inspriring	-	-	0
harmonic	-	-	-
light	-	-	0
clean	-	-	-
quality	0	-	-

Conclusions

There are two main take-aways from this study. As seen in Table 11.1 people tend to attribute complex cognitive constructs to apartments. Those are represented as factors. A building company can have an advantage when knowing which and how to address them. The factors could be highlighted to the customers in a better way.

A second important finding would be the insight that VR technology can aid to represent those factors in a more suitable way than the bare building plan. However, as seen in Table 11.2. none of the VR technologies can accurately convey the same feelings, emotions and attitudes as the real apartment can. Sure, the renderings can be improved, but the feeling is obviously not tied to factors such as contrast and colours but also to other aspects only the real product possesses.

Chapter 12: Shoes

By Shirley Coleman

This study was carried out as part of a three-year research project funded under European Commission Framework 5 to explore the application of Kansei Engineering in a European context. There were three academic partners providing input in statistics, design and biomechanics, and three small to medium enterprises (SMEs) specialising in three different types of shoe design (Bouchard et al, 2009).

Choice of domain

Clothing is a fundamental industry sector and shoes are an essential part of European dress. The study considered case studies in three different aspects of shoe ware.

In the initial case study, Kansei methodologies were employed to enhance the design of the designated product, "Men's Everyday Footwear." This investigation was conducted in collaboration with a fashion house manufacturer. This SME specializes in a diverse collection of shoes that undergo frequent updates. The company's focus was on determining whether Kansei Engineering could serve as a tool for crafting novel products with a strong likelihood of achieving success in the dynamic fashion market

In the second case study, Kansei techniques were used to investigate the design of specialist period footwear with products provided by a bespoke shoe design and craft SME. The designer of highly original fancy footwear creates each shoe customised and made-to-measure for a particular customer and purpose. The designer was interested in how the finished shoes were perceived as part of a continuous improvement cycle for the business.

The third case study delved into the realm of orthopedic footwear, conducted in collaboration with an expert orthopedic SME specializing in shoe manufacturing. In the domain of orthopedic footwear, a definitive medical purpose must be fulfilled while prioritizing utmost comfort. However, the challenge lies in ensuring that the products don't appear excessively "medical"; rather, they should resemble "regular" shoes and, in some instances, even be considered somewhat "stylish." The participants involved in the study were unaware that any of the shoes were orthopedic, and the medical aspect of some of the product sample was not used as a factor in the study.

Each case study held significance for the participating manufacturers as they aimed to ascertain how consumers perceive existing designs, evaluate the potential for predicting the impact of new designs, and determine the feasibility of establishing a set of "design rules." These rules would guide the integration of specific design attributes into new designs to evoke particular responses from the intended customers.

Case study 1 – Men's everyday shoes

Spanning the Semantic Space

The starting point of the Kansei data collection process was the identification of the relevant semantic universe of Kansei words and phrases associated with the target product. These terms were sourced from various outlets, encompassing street interviews, web pages, catalogs, and retail establishments. A comprehensive compilation of over 400 Kansei words was established for men's everyday footwear. Subsequently, this list was refined and condensed through the process of categorization, eliminating excessively intricate or ambiguous terminology for better manageability.

The words were then interpreted and grouped into 33 Semantic Differential Scales (Osgood et al 1957) as shown in Table 12.1. The semantic differential scales consist of two opposing adjectives or concepts.

Table 12.1: Kansei Words used in initial Semantic Scale Evaluation (van Lottum, Pearce and Coleman, 2006).

Tasteful -Tasteless	Impractical – Practical
Classic – Not Classic	Comfortable – Uncomfortable
Cushioning – Hard	Convenient – Inconvenient
Easy - Complicated	Timeless - Short lived
Original - Unoriginal	Cold – Warm
Exclusive - Common	Conservative – Outrageous
High Tech – Low Tech	Unattractive – Attractive
Good Quality – Bad Quality	Hard Wearing – Not Hard Wearing
Wide – Narrow	Fashionable – Unfashionable
Lazy – Active	Lightweight – Heavy weight
Retro – Futuristic	Hand Made – Mass produced
Contemporary – Traditional	Masculine – Feminine
Casual – Formal	Unstylish – Stylish
Leisure – Work	International – British
Young – Mature	Dependable – Undependable
Urban – Country	Cheap – Expensive
Fussy – Simple	

In the next stage consumers were asked to evaluate existing product designs using the Kansei words. The analysis was carried out separately in the UK, France and Spain, however, for simplicity just the UK version is described here.

A product sample of 27 real pairs of shoes was bought to be used for assessment. The sample was chosen to span as far as possible the design factors initially considered important by the researchers.

A small group of consumers (34 men between the ages of 25 – 65) attended product evaluation sessions for these 27 shoes. Respondents were given a shoe to handle and examine and were asked to mark where they considered the design lay on each semantic scale. In this study, we employed five-point scales, where each point on the scale was given a 'rating' ranging from 1 to 5. We then subjected the resulting scores from these semantic scales to various statistical methods. The aim was to identify inherent patterns within the semantic space and derive a set of seven factors. These factors could be utilized to evaluate a comprehensive range of product properties.

In summary, 400 words were reduced to 33 semantic pairs of words, and then these were further reduced to seven principal factors after being applied to a sample of real shoes.

Spanning the Space of Properties

A team of skilled footwear designers was tasked with pinpointing the specific physical design components that, in their view, would exert the most substantial impact on consumers' perceptions. An example of such a component is the material used for the upper part of the shoe. The designers were prompted to contemplate a scenario where consumers would form their judgments based solely on a product image, such as those found in catalogs or online stores. They were instructed to narrow down their selections to elements that would hold sway over consumers within this context. Given that footwear is intricate, with numerous pivotal design aspects, a large array of features initially emerged as potentially significant factors. After thorough discussion, the number of factors was reduced to a manageable list (Table 12.2). A subset of 100 shoe images was chosen from a collection of around 500 images. The selection aimed to encompass a diverse spectrum of designs.

Table 12.2: Design Elements and their associated categories (van Lottum, Pearce and Coleman, 2006).

Design Element Category	Setting	Setting Code
Upper Material	Leather	1
	Suede / Nubuck	2
Method of Application	Cemented/Moulded	1
	Welted	2
Upper Colour	Dark	1
	Light	2
Fastening	With Fastening	1
	Slip-On	2
Top Profile	Square	1
	Round	2
Sole / Heel Type	Flat sole, soft shape	1
	Angular shape heel	2
Sole Material	Leather	1
	Synthetic	2
Seal Distribution	None (plain toe)	1
	Moccasin	2
	Longitudinal	3
	Decorative toe cap	4
	Sport Style	5
Collar	None	1
	Reinforced	2
	Padded	3
Last Width	Wide	1
	Narrow	2

Within Table 12.2, the term "seal distribution" pertains to the arrangement of decorative seals on the shoe upper. Notably, "Sporty Seal" encompasses an extensive array of stitching patterns. "Sole / Heel Type" pertains to the general style of the sole. Footwear featuring a block heel falls under the category of "angular." Shoes without heels (such as those with a completely flat sole) or featuring a softly contoured sole are categorized as "Flat/Soft."

Synthesis

In the United Kingdom, around 44 individuals assessed between 30 and 100 shoes using the seven semantic scales specific to the UK. In sum, 59 shoes received adequate ratings from UK participants. This phase combines both Semantic Space and product design space, representing a central element of Kansei engineering.

Test of validity

The seven principal semantic space factors and the product design space were sense-checked to ensure they had face validity.

Model building

Kansei Engineering serves the purpose of establishing connections between tangible design components and subjective reactions from consumers. Ordinal logistic regression was employed to construct distinct models for every semantic scale. Here, the "semantic scale score" was treated as the dependent variable, while the encoded design elements were utilized as independent variables to predict the semantic scale scores.

The objectives of the regression were threefold: firstly, to pinpoint the design attributes that held the most significant influence on each semantic scale; secondly, to ascertain which levels of these crucial design attributes prompted a positive reaction on the semantic scale;

and thirdly, to determine the feasibility of generating a predictive model for each semantic dimension.

Summary of Results

Data quality

Data quality was enhanced by making sure instructions were clear and checking for errors on data input to spreadsheets via exploratory data analysis.

Semantic Space

The seven factors extracted from the raw semantic score data explained the % variance explained shown in Table 12.3:

Table 12.3: % variance explained by each of the 7 extracted factors (van Lottum, Pearce and Coleman, 2006).

Factor	1	2	3	4	5	6	7
% Variance	16.7	10.5	8.7	8.6	8.4	4.6	4.1

Factor 1 displayed a notable influence from terms like contemporary, young, classic, original, and timeless. This was construed as being associated with the visual appeal and originality of the design.

Factor 2 clustered words like attractive, stylish, and tasteful, indicating a connection with the overall attractiveness of the design.

Factor 3 included descriptors related to the shoe's quality and finishing.

Factor 4 grouped words connected to comfort.

Factor 5 related to practicality.

Factor 6 pertained to the convenience offered by the design.

Factor 7 encompassed solely the adjective pair "urban-country." However, it became evident that this scale had been misconstrued by some respondents. For this reason, Factor 7 was utilized cautiously in subsequent analyses.

Data modelling

The ordinal logistic regression models predicted the probability of a higher preference score in a semantic scale with respect to the design elements. The design element 'Seal Distribution' was found to be a significant factor for all of the semantic scales except the consumer perception as to whether the shoe was comfortable or easy. Further results are available in Bouchard et al (2009).

Discussion

There is always a risk of evaluator fatigue biasing the results. For this reason, the order of shoe presentation and order of semantic scales was varied at random between respondents.

Securing examples of shoes that encompass all necessary combinations of design factors proves challenging. Nevertheless, the advantage of a well-balanced image set lies in its capacity to evaluate the impact of each design element on various outcomes with consistent precision across each level of every design element. However, attaining a balanced product sample across all design elements is not always straightforward. Some combinations of elements might not naturally occur, like pink brogues, for instance. To address this, if needed, an approximately balanced subset can be assembled by choosing a subset of elements and limiting the levels to 2 or 3. To achieve such balance, a simulation was employed to generate a subset containing the most prevalent products.

The results showed how to design a shoe to obtain a high score on a chosen semantic scale. For example, choose Nubuck / Suede (vs.

Leather) to give a higher probability of a casual, fashionable or comfortable shoe.

The Kansei experiment was extensive and time consuming but gave significant results. The manufacturers were pleased with the outcomes and considered that they were now better able to design a balanced portfolio of shoe offerings.

Case study 2 – Bespoke fancy shoes

Spanning the Semantic Space

The small and skilled artisan business produces top-notch, vintage-style footwear suitable for both men and women. They excel in crafting precise historical reproductions and custom-designed shoes tailored for specific needs, such as recreating the footwear worn by Kublai Khan during his horseback journey across the Mongolian steppes.

The customer base for this specific product is smaller in comparison to men's everyday footwear. Thus, instead of seeking opinions from a broad cross-section of the population as was done for men's everyday footwear, the descriptive words were sourced from a different approach. These sources included the SME itself, various media outlets (such as websites, newspapers, magazines, and leaflets), and the members of the research consortium.

The Semantic Space encompassed approximately 200 words, which could be categorized into various broad descriptive themes: regal attributes (e.g. aristocratic, majestic, etc.), precision (e.g. flawless), historical attributes (e.g. period piece), general descriptive terms (e.g. adorable, amazing), design characteristics (e.g. comfortable, durable), exterior embellishments (e.g. adorned, elaborate), vividness (e.g.

eccentric, outlandish), enchantment (e.g. fairy tale), and national identity (e.g. British style). The comprehensive list of words was then distilled into the following 11 semantic scales:

- Well Made
- Luxuriant
- Enchanting
- Fancy
- Comfortable
- Ridiculous
- Sophisticated
- Attractive
- Authentic
- Tactile
- Unique

When it comes to reproducing historical footwear, maintaining authenticity (i.e., staying faithful to the original) is paramount. As such, the manufacturer is unable to alter the style of the products or the materials used (unlike in the case of everyday footwear). However, the utilization of the Kansei technique empowers us to identify which design elements are associated with perceptions of attractiveness, luxury, sophistication, and the like.

Spanning the Space of Properties

The product range is limited and devoid of standardized items; everything is custom-madee for individual customers. The manufacturer was requested to outline the product range, furnish examples, and provide a list of design elements. Consequently, a digital presentation was assembled, encompassing 14 product images. These images featured diverse styles, spanning a wide range of historical periods from 200-300 AD (Roman) to the present day. The styles varied greatly, ranging from highly ornate to quite plain. The upper materials exhibited different finishes: some were smooth, others had a pile, some featured punched holes, and some were a combination of fabric and leather. All 14 designs were utilized as the product sample.

To gauge the emotional response of each respondent to each image, a questionnaire was formulated. This questionnaire incorporated the application of semantic scales to every image within the product sample.

Gathering and surveying a customer base

The subsequent phase involves identifying the population to be surveyed regarding the footwear. A search of the internet and yellow pages yielded a selection of companies in the UK associated with stage attire and historical reproductions. This pool of potential respondents encompassed a diverse range of organizations: theater companies, period costume rental shops, period costume manufacturers, drama/theater/design schools, and museums. The customer base could be classified into three categories: 'producers of costume shoes,' 'users of costume shoes,' and 'trainers/educators in theater work and/or design.'

A total of 100 questionnaires were distributed to individuals considered 'experts' in the realm of historical reproduction footwear. These included 20 customers of the SME, forming a potential British respondent group. The response rate stood at approximately 25%, resulting in 26 companies participating. In addition, 16 non-expert volunteers were also willing to partake, some of whom had been involved in the evaluation of men's everyday footwear.

In the questionnaire, the list of semantic scales from the initial stage was randomized for each shoe, ensuring unbiased responses during questionnaire completion. Each respondent was asked to express their opinion on every word for each shoe using the Likert scale (1=strongly disagree to 5=strongly agree). Respondents were also queried about their satisfaction level if they had ordered the product for an appropriate purpose (e.g., Roman sandals for a biblical production).

Satisfaction opinions were also assessed using a Likert scale ranging from 1=totally dissatisfied to 5=totally satisfied.

The questionnaire initiated with introductory queries about the respondent's business involvement, such as historical costume hire/manufacture, fancy dress hire/manufacture, made-to-order clothing, retro clothing (e.g., 1960s, 1970s), costume design, museum display design, historical interpretation, military uniforms, live interpretation, or 'other.' Respondents were given the option to specify their line of work if it didn't align with the provided categories. They were also invited to share their age and gender, and they could choose to remain anonymous or disclose their business name. Additionally, they had the choice to opt-in for mention at upcoming marketing events, which served as an incentive for completing the questionnaire. Notably, it was observed that non-experts were more cooperative due to the more captivating shoe sample and the diverse product range compared to men's everyday footwear. The questionnaire underwent a preliminary test run with relevant organizations showing interest in such footwear, as well as with members of the public.

Due to the limited sample size and the intricate diversity among the shoe designs, the derivation of mathematical design rules linking design elements to 'emotional' attributes wasn't feasible. Instead, a qualitative analysis was conducted. This qualitative analysis followed a preceding quantitative statistical analysis of the semantic scores. The shoes were plotted in multidimensional space, and a qualitative assessment determined whether shoes that clustered together in terms of consumers' emotional sentiments shared similar characteristics. This hybrid approach was adopted to address the application of Kansei techniques in a new domain and within the context of a unique SME (Pearce and Coleman, 2008).

Synthesis

The numbers of completed questionnaires for each of the respondent types is given in Table 12.4.

Table 12.4: Breakdown of number of respondents taking part.

Full Set of Replies	**N=42**	
Gender*	Male=16	**Female=25**
Respondent Type	**Member of Public=16**	**Organisation dealing in Period Footwear=26**

* 1 respondent did not disclose their gender.

Model building

The first phase of the analysis centered on examining the respondents' perspectives regarding their opinions on the 11 semantic scales. Each of the 42 participants had evaluated 14 products, and the average rating for each product was computed across all the semantic scales. A higher mean value indicated greater agreement among subjects that the product possesses a particular attribute, such as being 'fancy.'

Moving to the second phase, the analysis focused on establishing the semantic axes through factor analysis. This process allows each footwear item to be represented in a multi-dimensional space, enabling the identification of similarities in respondents' perceptions.

Non-standardized principal component-based factor analysis was deemed suitable since each of the 11 emotional words, from 'attractive' to 'well made,' is measured using the same units (i.e., on the Likert Scale). The choice of non-standardized (covariance-based) analysis ensures that the inherent variability in the variables is effectively utilized.

A scree plot indicated that 4 factors should be extracted. The results for all respondents show that:

Factor 1 measures the extent to which the footwear is attractive, enchanting, fancy, luxuriant and sophisticated.

Factor 2 emphasises footwear which respondents feel is fancy, *ridiculous* and unique. Note that attractiveness is viewed oppositely.

Factor 3 emphasises footwear which is comfortable and tactile.

Factor 4 emphasises footwear which is authentic, unique and well made.

The scores for the products on the first two factors are displayed in 2D space in Figure 12.1.

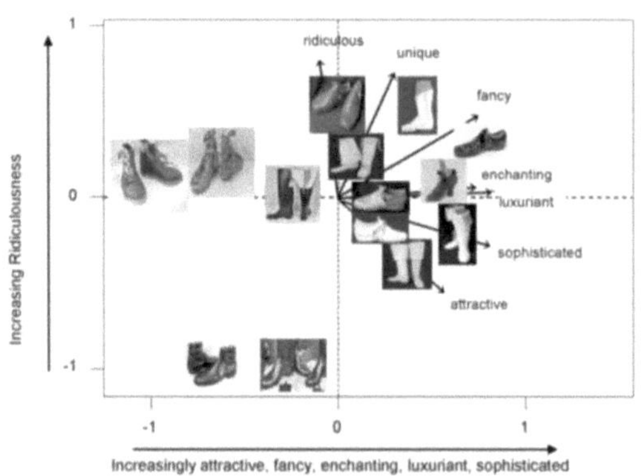

Figure 12.1: Plot of Factor 2 (y axis) versus factor 1 (x axis) with product illustration superimposed.

It can be seen that the products with the lowest score for ridiculousness in the lower left quadrant are the products used in the present day. The Roman sandals in the top left quadrant are considered to be rather ridiculous and also not very attractive. The long white boots in the lower right quadrant are considered rather enchanting. The cluster of shoes in the top right quadrant are all considered luxuriant but a bit ridiculous.

Upon reviewing the average responses for each shoe, noticeable distinctions emerged between age groups concerning their perceptions of the footwear's 'ridiculousness.' Younger individuals appeared considerably more inclined to label the footwear as ridiculous compared to older individuals, who either genuinely deemed the products ridiculous or were more reserved in expressing such views. Similarly, variations surfaced among age groups in relation to the term 'well made.' Generally, older respondents were less inclined than younger ones to perceive the products as 'well made.'

Furthermore, a noteworthy gender-based discrepancy emerged: males rated the footwear significantly higher than females in terms of 'ridiculousness.' This indicates that either males genuinely found the products more ridiculous compared to females, or they felt more comfortable expressing such opinions.

Interestingly, the general public's perception leaned toward viewing the products as more 'ridiculous' compared to organizations specializing in period footwear. This suggests that either members of the public found the products more ridiculous, or they felt more at ease expressing such sentiments. Additionally, the general public tended to view the products as more 'unique' compared to organizations that regularly encounter 'period footwear.'

Validation

After comprehensively analyzing all the outcomes, we can ascribe emotional labels to each piece of footwear. For instance, the high-heeled shoe, which is also the only ladies' shoe in the sample, emerged as the least comfortable and tactile product. Notably, this footwear is notably narrow and features the highest heel. Conversely, the items of footwear that closely resemble contemporary 'slipper' styles were deemed the most comfortable. These products are characterized by soft materials and a wider toe area. The long boots garnered associations with sophistication.

Discussion

The central emphasis of the case study was on the emotional reactions elicited by the footwear. The products are custom-made and predominantly consist of historical reproductions, so the design elements correspond to historical designs rather than being manufacturer-driven. Consequently, the manufacturer's principal interest was in understanding the perception of the products, such as how authentic the reproductions appeared. In this regard, the case study yielded positive results, affording the manufacturer improved insight into the customer base. This information could then be leveraged to refine the footwear even further in alignment with customer requirements.

Case study 3 – Orthopaedic shoes

The SME manufactures “off the shelf”, semi-customised and fully customised orthopedic shoes. The customer base includes elderly people and those with ailments such as rheumatoid arthritis and diabetes. A previous study by Williams and Nester in 2006 highlighted the differences in choices made between people with different ailments. They found that sufferers of rheumatoid arthritis are often in pain and look for a comfortable product whereas diabetics may have reduced sensitivity in their feet making them unaware of poor fitting shoes and tend to favour ‘style’ over comfort.

The design of orthopedic footwear is quite different from everyday products. Customers' reasons for buying/liking a product tend to be very different when compared to people who can just buy shoes and boots "off the shelf". Podiatrists are also well aware of patients' reactions to orthopedic products and thus are well qualified to give their opinion as to how they feel that patients will react to specific products.

Spanning the Semantic Space

Kansei Engineering initially involves looking for emotive words/terms associated with the product, these are then reduced to a smaller subset for a detailed investigation. For the study of orthopaedic footwear, appropriate sources such as the web, newspapers, magazines and various advertising mediums were searched to create the full set of emotive words for the set of shoes/bootees in question.

The set of words was then reduced using the views of experts and non-experts in the field. The following set of 11 words was chosen:

- High quality
- Smart
- Comfortable
- Attractive
- Boring
- Shapeless
- Suitable for the purpose
- Fashionable
- Looks Long-lasting
- Clinical
- Functional

The reduced set of words appear on the questionnaire as a set of semantic scales.

Spanning the Space of Properties

The research team worked closely with the manufacturer of the orthopaedic footwear to evaluate which design elements were influencing emotion. It was decided that a suitable sample contained products which (i) were diverse in style (ii) were interesting to the SME as regards their acceptability. Sixteen products were chosen for the study.

The design elements and their associated levels in Table 12.5 were taken into consideration when selecting the products:

Table 12.5: The Design Elements

Design Element	***Level***	
Material	Leather	Non Leather
Type of Shoe	Casual	Formal
Fastening	Tied	Velcro
Sole Height	Elevated	Low
Colour	Black	Non Black
Age Range	Older Wearer	Younger Wearer

For convenience, photographs of shoes (rather than real products) were used for evaluation and a mixture of male and female-styled products were chosen.

Synthesis

The set of semantic words appear in random order for each shoe/bootee picture. Each respondent is asked to cross the box which is closest to their feelings for each word.

The questionnaire also asks the respondents to state their age and sex. Therefore we are able to see how males/females view male/female products. Podiatrists are asked if they treat patients with rheumatoid arthritis and/or diabetes and the patient age range. If wearers are questioned, they are asked if they suffer from rheumatoid arthritis and/or diabetes.

Gathering respondents

Orthopaedic shoe/bootee wearers make up a sensitive area of the population. Most wearers are patients of some kind either (i) suffering from a specific ailment and/or (ii) are elderly. They are generally accessible only through their podiatrists, orthotists etc.

We sought the opinions of:

1. People who are wearers of orthopaedic shoes/bootees

2. Podiatrists who deal with wearers of orthopaedic footwear in their day-to-day lives.

We obtained general interest from various parties (foot surgeons, podiatrists, orthotists etc) throughout the UK, and these people were willing to distribute questionnaires to patients.

We obtained contacts through various means: personal contacts, various societies and associations. Before these individuals could distribute the questionnaires, however, Primary Care Trust/Acute Trust management approval is required in each location to ensure that the questions do not upset or offend patients.

Regarding ethical considerations, the study is regarded as a 'service evaluation' of the manufacturers' products. Because it is "motivated to define current care; defined to answer the question *what standard does it achieve*; measuring current service without reference to a standard and involving no more than administration of a questionnaire", it does not require ethical permission.

Questionnaires were mailed to individuals and the respondent sample is shown in Table 12.6.

Table 12.6: Breakdown of number of respondents taking part.

	Mild	*Moderate/Severe*
Female	***11***	***13***
Male	***4***	***5***
Not Known	***-***	***1***
TOTAL	***15***	***19***

Model building and validation

Non-parametric statistics (including the Wilcoxon Mann Whitney test), were used to evaluate the differences due to sex of respondent and ailment type. The sample size is low and statistical significance is hard to establish. However, the results can be illustrated graphically and various qualitative observations can be made. Such observations can be used to direct and justify further research.

The research unveiled that, in general, older individuals express greater enthusiasm toward orthopedic products compared to their younger counterparts. Patients aged over 60, on average, perceive these products as more attractive, less mundane, more comfortable, more functional, of higher quality, longer-lasting, more modern, smarter, and better suited for their intended purpose. Furthermore, in comparison to those below 60 years of age, the elderly respondents are more prone to believe that wearers of orthopedic products would be content using these items.

Figure 12.2 showcases some of the outcomes. It's evident that individuals aged over 60 would exhibit notably higher satisfaction levels with specific products (such as a laced-up taupe shoe (8), an open-toed sandal (9), and a blue leather lace-up (10)) when compared to younger individuals.

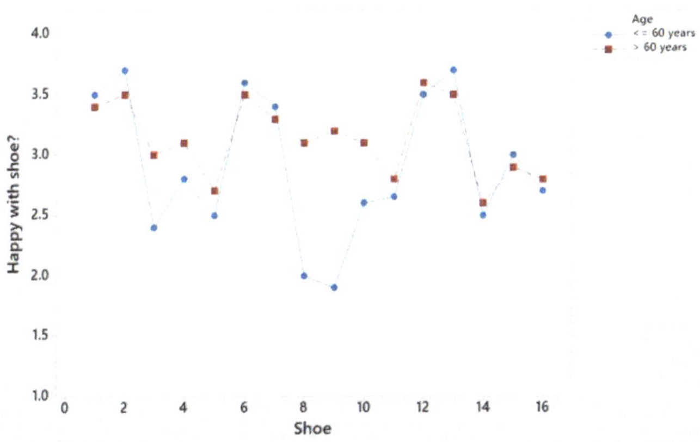

Figure 12.2: A comparison of product appeal by age.

Patients with mild conditions, who can easily find off-the-shelf footwear, typically perceive orthopedic footwear as more mundane, functional, and comfortable when compared to individuals with more severe conditions. Conversely, those individuals with more severe foot conditions find these products to be more attractive, of higher quality, and modern (in terms of aesthetics), albeit less comfortable than those with milder conditions.

Individuals dealing with moderate to severe conditions generally believe that the products would be less comfortable than those with mild conditions, as depicted in Figure 12.3. Specifically, training shoes (products 1 and 13) garner notably high comfort ratings from individuals with mild conditions. Interestingly, the black training shoe (product 13) is highly regarded for comfort by those with moderate/severe conditions, whereas the tan and cream training shoe (product 1) receives substantially lower ratings from the same group. Evidently, color is influencing the responses.

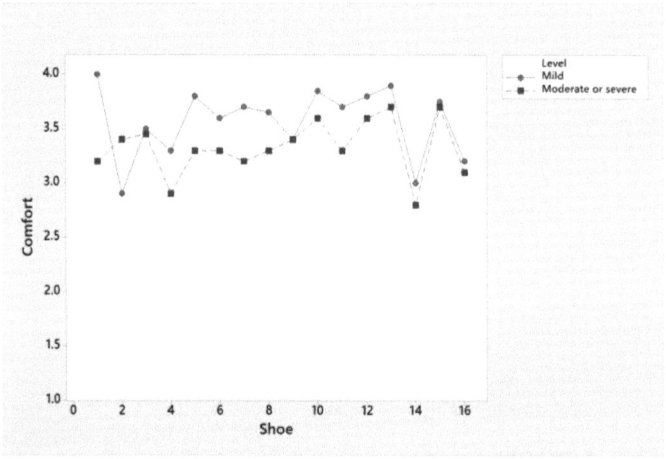

Figure 12.3: A comparison of product comfort by condition.

The interaction plot depicted in Figure 12.4 reveals that patients over the age of 60, who are also dealing with moderate/severe conditions, tend to hold a significantly more favorable overall view of the products in comparison to individuals under 60 with similar moderate/severe conditions.

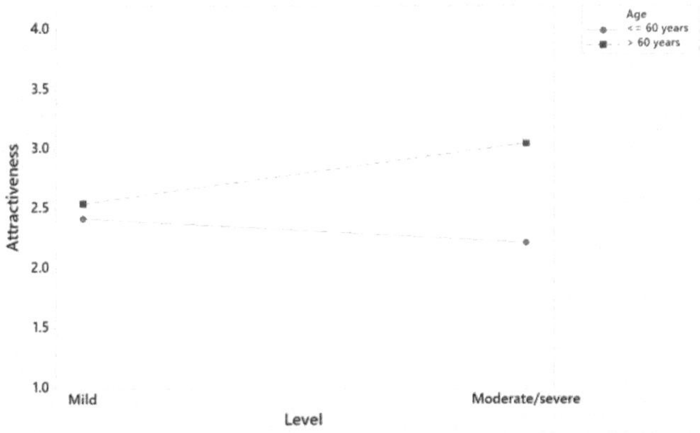

Figure 12.4: Interaction plot of product attractiveness by age and condition.

Those patients who are younger than 60 with a mild condition generally favour the products more than those who are younger than 60 with a moderate severe condition.

Some semantic scales showed a clear difference between the views of females and males. Figure 12.5 shows that females rated the shoes more long-lasting than males. Both sexes agreed that Product 9 (a lady's cream leather open-toed sandal) is probably not long lasting.

The plots in Figures 12.2 to 12.6 show how the Kansei data can be graphically analysed. As said already, the sample of respondents is small and the results may not be repeatable. However, there is considerable consistency in the results and the fact that they "make sense" implies a sound level of validity.

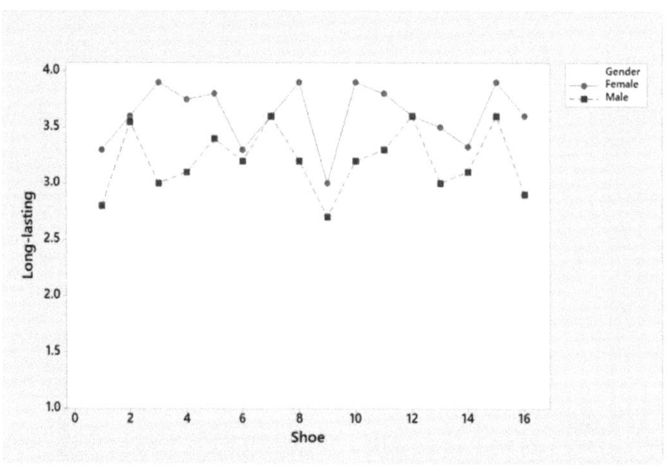

Figure 12.5: A comparison of results for "long-lasting" by gender.

Discussion

The research was highly valued by the SME. The main finding was the surprising result that orthopaedic shoe wearers really do rate the appearance of the shoes as well as comfort, value for money and other factors. The research reinforced the requirement for the manufacturer to continue to produce a wide range of shoes and to choose appealing colours and designs as far as possible, and to pay attention to changing fashions.

Chapter 13: Website Emotion

By Anitawati Mohd Lokman

Kansei Engineering is an approach to product design that takes into account users' emotional responses to a product. This approach has been applied to website design in order to create websites that are not only functional, but also emotionally satisfying for users.

In the context of website design, KE focuses on understanding the emotional responses of users towards various design elements such as colour, texture, layout, and typography. The emotional elements of website design are critical in influencing user experience and satisfaction. By applying KE in web design, designers can create visually appealing and user-friendly websites that improve user engagement and retention.

The application of Kansei engineering in web design involves understanding the emotional needs of the users and incorporating them into the design parameters of the website. By doing so, designers can create a website that not only meets technical requirements but also satisfies the emotional needs of the users, leading to increased user engagement and satisfaction.

Several studies have been done investigating the aspect of emotion in website design by the use of Kansei Engineering. Among others are the work on web interface design for academic institutions by Zhou et al. (2022), Kansei Web Design Guideline by Lokman (2010), Kansei evaluation in e-Learning (Hadiana & Lokman, 2016), comfortability in news website (Bidin & Lokman, 2018), Tourism mobile apps (Wozzari & Lokman, 2018), and many more.

In this book, we introduce the work by Lokman on the development of Kansei Web Design Guideline (KWDG). The case study in clothing e-Commerce website design introduced KE adoption in website design by applying the Kansei Product Design Framework (Refer to Figure 13.1) to build Kansei website, demonstrating Kansei measurements and empirical analysis. The study discovers the Semantic Space of Kansei words and the relationships between Kansei responses and website design. The website domain presented in this section is clothing e-Commerce (e-Clothing) website user interface design (UID).

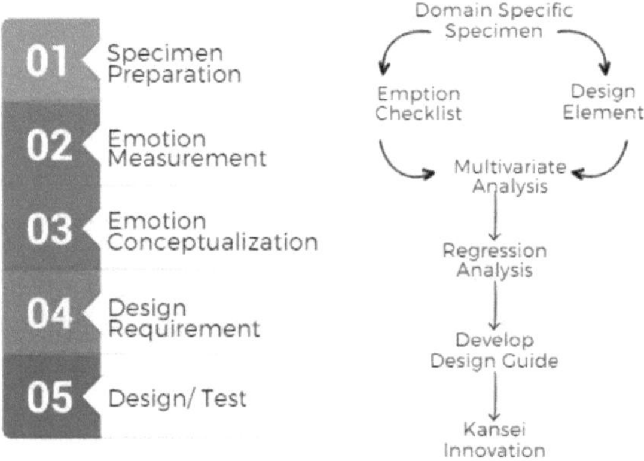

Figure 13.1: The Kansei Product Design Framework (Lokman, 2010).

The Kansei Website Design Framework is a process that involves incorporating specific emotional concepts into the design of a website in order to create a positive user experience and increase the website's success. The emotional concepts, such as comfortability or aesthetic sensation, are derived from Kansei Engineering, which aims to capture and bring customers' psychological feelings into the design of products or systems. Incorporating Kansei Engineering into website design is becoming increasingly important, as research highlights the importance of considering human Kansei in the development of technological and everyday life products. By creating a Kansei website, businesses and organizations can provide a more personalized and engaging user experience, which can lead to increased user satisfaction and ultimately drive website success. The implementation of the framework in the case of e-Clothing is described below:

Specimen Preparation

Firstly, one hundred and sixty-three e-clothing websites were selected based on their visible differences in design, e.g. page colour, font size, menu shape, presence of images, etc. All these websites were then analysed to identify the individual design elements of the website and categorised into 'attributes' and 'values'. The term 'attribute' in the research refers to feature elements in all the examples. For example, the background colour, the way the product is presented, the position of the menu, etc. On the other hand, 'value' refers to the values of the individual elements. For example, white, green, blue, or image, video, animation, etc.

The reason for this categorisation into attributes and values is to maintain the perspective of design from a human point of view. The reason is that a lay person sees design in many different ways. Some of them focus on certain elements, others look at design as a whole. The translation of design elements, focusing only on the designer's point of view, leads to a one-sided definition of designs, whereas the classification of designs from a human point of view is the know-how in KE. In the analysis phase, this classification is used to find out which consumer's Kansei is highly associated with which item and, in particular, corresponds to which category, so that the design element of each Kansei can be determined.

A number of controls were applied in the selection of the websites. This enabled the study to identify detailed design elements on each site, thus enabling analysis of differences and similarities in terms of adherence to design rules. The selected specimens were further analysed to identify valid websites for evaluation, eventually allowing thirty-five specimens to be used. The selected specimens were websites selling clothing for teenagers. The specimens are numerically coded from one to thirty-five and the snapshots of the specimens are shown in Figure 13.2.

Figure 13.2: Examples of Website Specimens.

Next, Kansei words were selected. The use of Kansei words in website design is important to create a positive user experience and contribute to the website's success. In this study, Kansei words were identified using web design guides, experts, and relevant literature. Forty Kansei words were selected based on their suitability to describe a website. These words include adorable, professional, impressive, and more.

The selected Kansei words were used to develop a website evaluation checklist, which was organized into a 5-point semantic differential scale (SD). This approach is similar to the psychological scaling method in Kansei Engineering, which involves creating a Likert-type scale using Kansei words. By using Kansei words and the developed checklist, designers can evaluate their website's emotional impact on users and make necessary changes to improve the user experience.

The Kansei Measurement

Recruiting the appropriate participants for a research study is crucial to ensure the study's success and reliability. In this case, the study focuses on the Kansei of the target consumers, which are youths and online shoppers, for a product design in e-clothing. Therefore, it is essential to recruit these specific target consumers as respondents for the study.

The study recruited 120 online shoppers among the youth as subjects for the evaluation. To ensure the evaluation's accuracy, the subjects were systematically shown thirty-five valid specimens in a controlled manner and asked to record their impressions on the checklist using the scale provided. The checklist was developed using Kansei words that were synthesised from web design guides, experts and relevant literature and organised into a 5-point semantic differential scale (SD). Before the assessment began, the subjects were given briefings about the rules, the tools, and the aim of the measurement to ensure their accuracy in the assessment. The subjects were also given time to familiarise themselves with the specimens before the assessment began. The order of the Kansei words in the checklist was changed to avoid bias, and the subjects were given approximately three minutes to rate their feelings towards each specimen and a break to relax and clear their minds. The whole session takes about two hours.

The Emotion Conceptualization

Multivariate analysis is a statistical method used to analyse the relationship between multiple variables at once. It is a powerful tool for understanding complex data sets and identifying patterns and relationships between variables. In the context of emotions, Multivariate analysis can help to identify the emotions that arise from a set of responses, and how they are related to each other, thus can provide a more comprehensive understanding of the structure of emotions.

In this case study, Multivariate Analysis was used to comprehend the concept of emotions that form from the responses. Two descriptive multivariate techniques, namely Principal Component Analysis (PCA) and Factor Analysis, were implemented with the average Kansei responses dataset. These analyses helped the research conclude the concept of emotion in Website UID.

Understanding the Kansei Semantic Space

The primary outcome of implementing Kansei Engineering is to identify the emotional responses that a product evokes in users and incorporate these responses into the design process. The Semantic Space of Kansei words is used to describe the emotional responses in a certain vector space defined by the semantic expressions. The Semantic Space for the e-Clothing website was analysed through PCA.

Figure 13.3 depicts the PC loadings for the first and second principal components based on the evaluation result. The PC loadings show how much the Kansei evaluation affects the variables used to determine the semantic structure of KW. Specifically, the words "old-fashioned" and "boring" had low loading values along the first principal component, implying the axis of attractiveness. In contrast, the words "gorgeous," "impressive," "stylish," and "appealing" received the highest loadings, indicating high attractiveness.

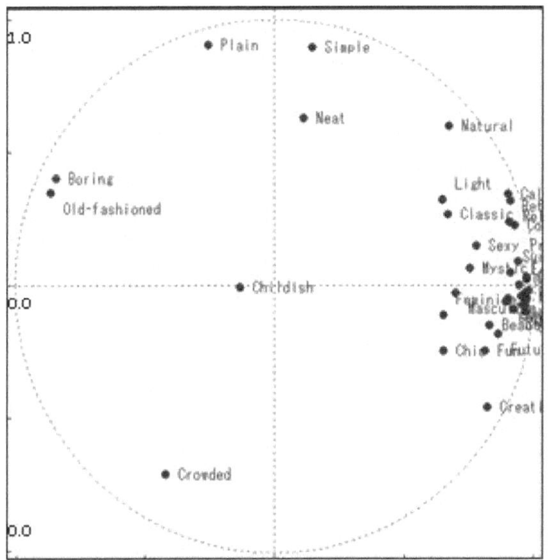

Figure 13.3: PC Loading for the e-Clothing Website Design.

Additionally, the words "plain" and "simple" had the highest loading values along the second principal component, implying the axis of complexity, followed by "neat" and "natural." The word with the highest negative value was "crowded." Therefore, it is possible to conclude that the Kansei structure in website design has two components: attractiveness and simplicity, and the blending and balancing of these two components are important determinants of new website design.

Figure 13.3 shows the PC loadings for the first and second principal components based on the Kansei evaluation result, and the words with high loadings indicate high attractiveness and simplicity, respectively. Therefore, a new website design should balance and blend these two components to create an appealing and simple design.

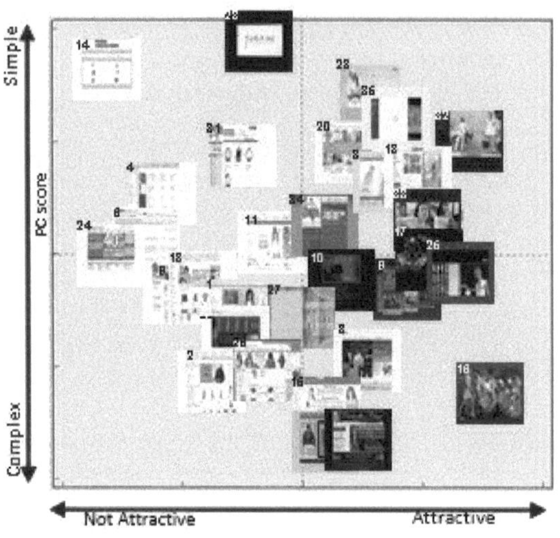

Figure 13.4: Principal Component Score for the e-Clothing Website Design.

The principal scores of the specimens were analysed and plotted in Figure 13.4 to identify their corresponding Kansei space in this study. It was found that strong meanings were held by websites located at the edges of the Kansei space. Specifically, website no. 24, located at the very left edge, was rated as "not attractive" with very low scores. Similarly, websites no. 16, 25, and 32, located at the very right edge, were rated as "very attractive". Websites no. 30 and 12, located at the bottom edge space, were rated as "crowded".

It was observed that websites in the "not attractive" category, such as sample no. 24 and 14, had small-sized pictures, mostly text, observable empty spaces, and lacked any modelling on clothing. In contrast, "Attractive" websites, such as sample no. 16, 32, and 25, had large-sized pictures, very few empty spaces, less text, and used models to showcase their clothing. It was also noted that websites with darker backgrounds tended to be categorized as "attractive".

Understanding the Emotional Structure

Factor Analysis (FA) is a statistical technique used to identify the underlying structure or dimensionality of observed data and reveal the latent constructs that give rise to observed phenomena. The technique identifies and examines clusters of inter-correlated variables, called factors or latent variables, to better understand the data.

In the study of website emotion, Factor Analysis (FA) was used to identify significant emotional factors that can help determine the concept of emotion in Website UID. The results of the analysis are used to refine the outcome of PCA. Table 13.1 displays the FA after varimax rotation. Varimax rotation is a popular method developed by Kaiser (1958) used to simplify the interpretation of variables in factor analysis.

Table 13.1: Factor Contribution.

Factors	Variance	Contribution	Cumulative Contribution
Factor1	16.09262	40.23%	40.23%
Factor2	12.29421	30.74%	70.97%
Factor3	3.427578	8.57%	79.54%
Factor4	1.856272	4.64%	84.18%
Factor5	1.810882	4.53%	88.70%
Factor6	0.923415	2.31%	91.01%
Factor7	0.370649	0.93%	91.94%
Factor8	0.250962	0.63%	92.57%

As shown in Table 13.1, the first factor explains 40.23% of the data and the second factor explains 30.74% of the data, indicating that both factors have a significant influence on emotions. Together, the first two factors represent 70.97% of the variability, while including the third factor would increase the proportion of variability explained to 79.54%. The fourth factor and any additional factors only explain a minimal amount of variability (less than 5%) and can be considered insignificant.

Table 13.2: Factor Loadings for Emotions.

Variable	Factor1	Variable	Factor2	Variable	Factor3	Variable	Factor4
Old-fashion	-0.6365	Boring	-0.7165	Crowded	-0.75621	Old-fashion	-0.32332
Boring	-0.4849	Old-fashion	-0.56005	Creative	-0.38735	Classic	-0.31312
Professional	0.80580	Pretty	0.68945	Sexy	0.27292	Impressive	0.20291
Cool	0.81133	Lovely	0.69002	Classic	0.27544	Masculine	0.21136
Gorgeous	0.81275	Elegant	0.70341	Boring	0.30859	Adorable	0.21181
Impressive	0.82273	Adorable	0.71303	Light	0.31383	Cool	0.28423
Surreal	0.84644	Charming	0.76368	Neat	0.31928	Interesting	0.30814
Sophisticated	0.84842	Sexy	0.78761	Calm	0.33916	Comfortable	0.32044
Luxury	0.87883	Cute	0.79405	Relaxing	0.34851	Lively	0.32850
Masculine	0.89911	Beautiful	0.81695	Natural	0.42488	Refreshing	0.39094
Futuristic	0.91316	Chic	0.93916	Plain	0.83900	Fun	0.49998
Mystic	0.94185	Feminine	0.94870	Simple	0.9241	Light	0.61059

Table 13.2 shows instance of factor loadings for emotions. The factor loading shows that the emotion of the website is structured into five factors; Exclusiveness, Gracefulness, Easiness, Lightness and Orderliness. These five factors altogether explains 88.70% of the total data, and thus influence highly to the design of website that embeds target emotion.

The first factor includes "Mystic," "Futuristic," "Masculine," "Luxury," "Sophisticated," "Surreal," "Impressive," "Gorgeous," "Cool," and "Professional." This factor is labelled as the "Exclusiveness" concept.

The second factor comprises "Feminine," "Chic," "Beautiful," "Cute," "Sexy," "Charming," "Adorable," and "Elegant," which are labelled as the "Gracefulness" concept.

The third factor consists of "Simple" and "Plain" and is labelled as the "Easiness" concept.

The fourth factor includes "Light" and is labelled as the "Lightness" concept.

The fifth factor comprises "Neat" and "Natural," which is labelled as the "Orderliness" concept.

The results indicate that the first and second factors, Exclusiveness and Gracefulness, account for 70.97% of the variance in the data, emphasizing their importance in website design. Therefore, it is recommended that designers incorporate these concepts to achieve optimal results. While Easiness, Lightness, and Orderliness also contribute to the overall emotional experience, their influence is relatively weak. As a result, these concepts should be used as supporting elements in website design with the target emotion in mind.

The Design Requirement

The study employed Partial Least Squares (PLS) analysis to uncover relationships between the Kansei (emotional responses) and design elements (x variables). PLS analysis was used to identify the impact of design elements on each Kansei, determine the best and least fitting values for each design element, and understand how different samples evoke specific Kansei responses. This choice was motivated by the capacity of PLS analysis to effectively manage a large number of x variables and numerous y variables.

The preliminary investigation of design elements in the earlier research phase resulted in the identification of 77 design elements and 249 corresponding values. For the purpose of PLS analysis, all these elements were transformed into dummy variables.

Table 13.3 displays a subset of the coefficient scores computed through PLS analysis. These scores were examined in the research to establish connections between emotions and design elements. The subsequent sections elaborate on how the utilization of these scores facilitates the identification of how various combinations of design elements influence emotional responses.

The study utilized PLS analysis to explore the relationship between design elements (x) and Kansei (y), with a focus on identifying the influence of each design element on specific Kansei, as well as the best and worst fit for each design element and the impact of each sample on the elicited Kansei responses. PLS analysis was chosen for its ability to handle a large number of x variables and tens of y variables.

In the earlier phase of the research, 77 design elements and 249 values were identified and converted into dummy variables for PLS analysis. The PLS coefficient score was calculated and analyzed to explore the relationship between design elements and Kansei. This allowed the researchers to identify the impact of design element combinations on eliciting specific Kansei responses.

The use of PLS analysis to analyze the relationship between design elements and Kansei can provide insights into the combinations of design elements that influence specific emotional responses. This methodology can be applied to a variety of product design and development contexts to ensure that products meet the needs and desires of users.

Table 13.3: Sample of the PLS Coefficient Score.

Design Element	Emotion				
	Adorable	Appealing	Boring	Calm	Charming
BodyBgColour-White	-0.0366	-0.0370	0.0245	-0.0253	-0.0355
BodyBgColour-Black	0.0065	0.0110	-0.0027	0.0284	0.00530
BodyBgColour-DkBrown	0.0604	0.0670	-0.0346	0.0345	0.0621
BodyBgColour-LtBrown	0.0132	0.0116	0.0060	0.0178	0.0211
BodyBgColour-Grey	0.0293	0.0365	-0.0440	0.0063	0.0340
BodyBgColour-LtBlue	0.0214	0.0042	-0.0116	-0.0121	0.0053
PageMenuShape-Curve	0.0006	-0.0106	-0.0079	-0.0070	-0.0117
PageMenuShape-Sharp	-0.0122	-0.0020	0.0218	0.0062	-0.0021
PageMenuShape-Mix	0.0287	0.0241	-0.0390	-0.0026	0.0264
PageStyle-Frame	0.0340	0.0254	-0.0396	0.0052	0.0055
PageStyle-Table	-0.0420	-0.0351	0.0420	-0.0181	-0.0112
PageOrientation-BC	-0.0447	-0.0416	0.0263	-0.0330	-0.0298
PageOrientation-Content	0.0327	0.0373	0.0111	0.0247	0.0347
PageOrientation-Header	-0.0546	-0.058	0.0374	-0.0345	-0.0623
PageOrientation-HF	0.0158	0.0153	-0.0181	0.0077	0.0147
PageOrientation-HSplit	0.0230	0.0217	-0.0185	0.0166	0.0221
PageOrientation-VSplit	0.0190	0.0341	-0.0233	0.0169	0.0288
PageOrientation-Plain	0.0242	0.0191	-0.0144	0.0153	0.0055
DominantItem-Pict	0.0467	0.0480	-0.0436	0.0147	0.0508
DominantItem-Adv.	-0.0297	-0.0323	0.0194	-0.0074	-0.0209
DominantItem-Text	-0.0561	-0.0455	0.0502	-0.0028	-0.0429

PLS Range for each emotion was calculated to determine the influence of design elements on emotion. Range calculation enables the identification of design influence, good and bad design. The maximum and minimum values were used to calculate the range, where

Range = PLSMax - |PLSMin|

Mean of Range was calculated, where

$$\overline{Range} = \frac{1}{n}\sum_{i=1}^{n} Range_i$$

Each emotion has a mean range, and if the mean value of a design element is greater than the average range, the item is thought to have a positive influence on design. A value range for each design element that is greater than the average value implies the best fit value, which heavily influences user emotion in Website UID.

Table 13.4 shows a sample of the results of influential design elements that have a range greater than the average range in each emotion. To illustrate the sequence of dominant design elements for each emotion, the results are sorted in descending order.

Table 13.4: A sample of the influence of design elements to emotion.

	Adorable		Appealing	
No.	Design Element	Range	Design Element	Range
1	Page Colour	0.11488	Header Bg Colour	0.12338
2	Product Display Style	0.10644	Face Expression	0.12216
3	Header Menu Bg Colour	0.10612	Header Menu Bg Colour	0.12077
4	Left Menu Font Colour	0.10370	Product Display Style	0.10646
5	Header Bg Colour	0.10218	Body Bg Colour	0.10574
6	Face Expression	0.10024	Page Colour	0.10091
7	Body Bg Colour	0.10015	Left Menu Font Colour	0.10085
8	Dominant Item	0.09980	Picture Style	0.09771
9	Header Font Size	0.09651	Page Orientation	0.09182
10	Main Text Existence	0.08813	Dominant Item	0.09141
11	Main Bg Colour	0.08587	Main Text Existence	0.08811

The result suggests that in order to design an 'Adorable' website, the designer must prioritise design elements based on their high influence, such as 'Page Color,' 'Product Display Style,' 'Header Menu Background Color,' 'Left Menu Font Color,' and so on. To design an 'Appealing' website, on the other hand, the designer must prioritise design elements based on their high influence, such as 'Header Background Color,' 'Face Expression,' 'Header Menu Background Color,' and so on. The complete set of analysis enabled the research to develop a design guide for Kansei Website, named Kansei Web Design Guideline (KWDG).

i) Designing the Kansei Website

To create a successful product that elicits the desired emotion, the designer should combine two or more emotional elements drawn from various emotion concepts. However, in this study, prototypes were developed based on individually selected emotional elements for confirmation purposes. This was due in part to the large number of design elements, and it is nearly impossible to combine emotional concepts. The five emotional elements were chosen from the KWDG (Refer to Table 13.5).

Table 13.5: Instance of the selected emotion and the design elements from the guideline.

Design Element	**Emotion**				
	Cute	**Feminine**	**Luxury**	**Masculine**	**Simple**
Body Bg Colour	Light Blue	Light Blue	Black	Black	Dark Brown
Body Bg Style	Texture	Texture	Colour Tone	Colour Tone	Picture
Page Shape	Sharp	Sharp	Sharp	N/S	Sharp
Page Menu Shape	Mix	Mix	Sharp	Sharp	Sharp
Page Style	None	None	None	None	None
Page Orientation	Footer	Footer	Vertical Split	Header	Content
Dominant Item	Picture	Picture	Picture	Picture	N/S
Page Colour	Grey	Pink	Black	Blue	Brown
Page Size	Small	Small	Small	Medium	Medium
Other Images?	Animal	Animal	Animal	Animal	Kids
Product Display Style	Filmstrip	Filmstrip	Filmstrip	Filmstrip	Catalogue
Product Try On	Yes	Yes	Yes	Yes	Yes
Product view angle	Rear	Rear	Mix	Side	None
Artistic Menu?	Yes	Yes	Yes	Yes	No
Empty Space?	Less	Less	Less	Less	More
Discount Ad. Existence	No	No	No	Yes	Yes

The guidelines were carefully followed in the creation of the prototypes based on the five targeted emotional concepts. Figure 13.5 shows a snapshot of the developed Kansei Website UID. It should be noted that, in the ideal case, the guideline should be used in conjunction with the designer's creativity to achieve the best results. The prototype went through a confirmatory evaluation phase, and validated.

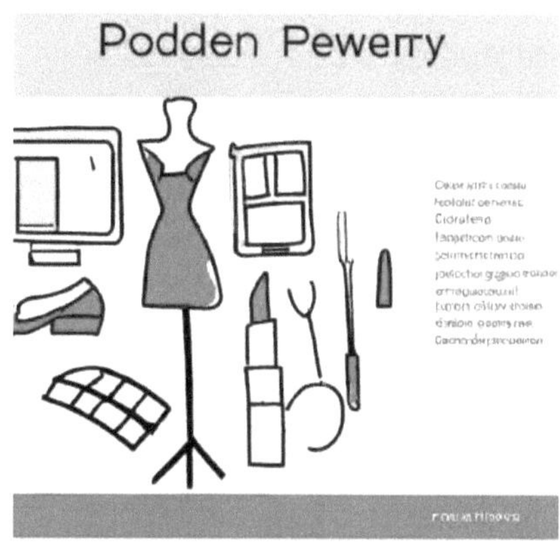

Figure 13.5: Example of the Kansei Website design (compare also: Lokman, 2010).

Discussion

The study applied the Kansei Website Design Framework to clothing e-Commerce website to demonstrate Kansei measurements and empirical analysis. The framework enables the discovery of emotional concepts to improve user experience in website designs. This study examines the Kansei of youth and online shoppers for e-clothing website design. The Kansei structure in website design shows two components: attractiveness and simplicity, and the blending and balancing of these two components are important determinants of new website design. The study utilised PLS analysis to explore the relationship between design elements and Kansei, with a focus on identifying the influence of each design element on specific Kansei, as well as the best and worst fit for each design element and the impact of each sample on the elicited Kansei responses. The results were used to develop KWDG, and prototypes were successfully developed based on the guideline. Future research could use the KWDG to design emotional evocative website, and replicate use the Kansei Web Design Framework in future studies.

KWDG, or the Kansei Engineering Web Design Guide, is a framework that provides guidelines for designing websites that elicit specific emotional responses from users. On the other hand, the Kansei Web Design Framework is a method for designing websites that incorporates the user's emotional responses into the design process.

Using the KWDG to design emotional evocative websites could lead to more engaging and memorable user experiences. Future research could use the KWDG to design websites that target specific emotional responses, such as happiness, excitement, or relaxation, and evaluate the effectiveness of these designs in eliciting those emotions from users. Also, replicating the use of the Kansei Web Design Framework in future studies could further refine the framework and identify best practices for designing emotionally engaging websites.

Overall, the use of the KWDG and the Kansei Web Design Framework could have significant implications for the field of web design, particularly in terms of creating more personalized and user-centered experiences. Further research in this area could contribute to a better understanding of the relationship between website design and user emotions, and ultimately lead to more effective and enjoyable online experiences.

References for Part 3

Arnold, K. (2002). Towards increased customer satisfaction. In *IKP/KMT*. Linköping University.

Bergman, B., & Klefsjö, B. (2002). *Kvalitet i alla led*. Studentlitteratur.

Bergqvist, I., & Domeij, O. (2001). Hemkänsla i en virtuell verklighet. In *EKI*. Linköping University.

Bouchard, c., Mantelet, F., Aoussat, A., Solvers, C., Gnzalse, J., Pearce, K., Lottum, D., Coleman, S., (2009), A European Emotional Investigation in the Field of Shoe Design, Product Development 7 (1-2) https://doi.org/10.1504/IJPD.2009.022274

Camargo, F. R., & Henson, B. (2015). Beyond usability: designing for consumers' product experience using the Rasch model. *Journal of Engineering Design*, *26*(4–6), 121–139. https://doi.org/10.1080/09544828.2015.1034254

Falk, J., Fredriksson, T., Wallin, R., & Nguyen, N.-A. (2016). *Developing a Box Design for Ice Cream*.

Friberg, E. (2010). Varumärkesstudie inom konfektyrbranschen- En kartläggning av Sportlunchs konsumenter. In *Department of Management and Engineering: Vol. Master*. Linköpings Unversitet.

Grzechnik, D., & Priithiviraj, A. (2010). A Kansei Engineering study on wafer based choclates. In *Department of Management and Engineering: Vol. Masters*. Linköping University.

Hair, J. F., Anderson, R. E., Tatham, R. L., & Black, W. C. (1995). *Multivariate data analysis with readings*. Prentice-Hall.

Ishihara, S., Ishihara, K., & Nagamachi, M. (1998). Hierarchical Kansei analysis of beer can using neural network. *Proceedings of Human Factors in Organizational Design and Management - VI*, 421–425.

Ishihara, S., Ishihara, K., Nagamachi, M., & Matsubara, Y. (1995). arboART: ART based hierachical clustering and its application to questionnaire data analysis. *IEEE Conference on Neural Networks*.

Kanda, T. (2002). A Method to evaluate Human Meal Kansei. *Internaional- International Journal of Kansei*, *3*(Kanda, T.), 13–20.

Kansei Enginering Software- KESo. (2017). Linköping University. www.Kanseiengineering.net

Komazawa, T., & Hayashi, C. (1976). *A statistical method for quantification of categorical data and its applications to medical science* (F. T. de Dombal & F. Gremy (Eds.)). North-Holland Publishing Company.

Levy, P, Yamanaka, T. (2006). Kansei information approach for an interdisciplinary design method proposal based on intuition. *9th International Design Conference DESIGN 2006*, 1475–1482.

Lokman, A. M. (2010). *DESIGN & EMOTION: The Kansei Engineering Methodology 1*(1), 1–14.

Lundblad, F., Raujol, B., Rapp, J., & Wester, J. (2022). MX Master Hologram Edition. In *Student course project poster TMKT71*. Linköping University.

Lundvik, A., Quere, T. L., Mir, P., & Welander, T. (2021). *Headphones for kids age 5-6*.

Marco-Almagro, L. (2011a). *Statisical Methods in Kansei Engineering Studies*. UPC Barcelona Tech.

Marco-Almagro, L. (2011b). *Statisical Methods in Kansei Engineering Studies* [UPC BarcelonaTech]. http://hdl.handle.net/10803/85059

Miguel, N., Valverde, A., & Schütte, S. (2022). ACHIEVING CUSTOMER ACCEPTANCE OF NOVEL PRODUCT FEATURES BY OFFERING CUSTOMER DELIGHT. *9th International Conference on Kansei Engineering- KEER 2022,*.

Mori, N. (2002). Rough set approach to product design solution for the purposed “Kansei.” *The Science of Design Bulletin of the Japanese Society of Kansei Engineering*, *48*(9), 85–94.

Nagamachi, M., Lokman, A. M. (1995). *Kansei Innovation: Practical Design Applications for Product and Service Development*. Taylor & Francis Group: CRC PRess.

Nagamachi, M. (1995). Kansei Engineering: A new ergonomic consumer-oriented technology for product development. *International Journal of Industrial Ergonomics*, *15*, 3–11.

Nagamachi, M. (1997). Kansei Engineering: The framework and methods. In M. Nagamachi (Ed.), *Kansei Engineering 1*. Kaibundo Publishing Co., Ltd.

Nagamachi, M. (2011). *Kansei/Affective Engineering* (M. Nagamachi (Ed.); 1st ed.). CRC Press.

Nagamachi, M., & Lokman, A. M. (2015). *Kansei Innovation: Practical Design Applications for Product and Service Development*. Taylor & Francis Group: CRC Press.

Nagamachi, M., & Mohd Lokman, A. (2015). *Kansei Innovation*. CRC Press LLC.

Nagamachi, M., Nishino, T., & Ishihara, S. (2001). Structural Analysis on Relations between Kansei Words. In M. G. Helander, H. M. Khalid, & M. P. Tham (Eds.), *The International Conference on Affective Human Factors Design*. Asean Academic Press, London.

Nagasawa, S. (1997). Kansei evaluation using fuzzy structural modelling. In M. Nagamachi (Ed.), *Kansei Engineering 1* (pp. 119–125). Kaibundo Publishing Co., Ltd.

Nagasawa, S. (2002). Kansei and business. *Kansei Engineering International- International Journal of Kansei Engineering, 3*(3), 2–12.

Nishino, T., Nagamachi, M., & Ishihara, S. (2001). Rough set analysis on Kansei evaluation of color. In M. G. Helander, H. M. Khalid, & M. P. Tham (Eds.), *The International Conference on Affective Human Factors Design*. Asean Academic Press.

Oluwafemi, S. A., & Yamanaka, T. (2014). Kansei as a Function of Aesthetic Experience in Product Design. In *Industrial Applicaions of Affecctive Engineering* (pp. 83–95). Springer.

Osgood, C. E. (1969). The Nature and Measurement of Meaning. In C. E. Osgood & J. G. Snider (Eds.), *Semantic Differential Technique - a Source Book* (pp. 3–41). Aldine publishing company.

Osgood, C. E., & Suci, G. J. (1969). Factor analysis of meaning. In C. E. Osgood & J. G. Snider (Eds.), *Semantic differential technique - a source book* (pp. 42–55). Aldine Publishing Company.

Pearce, K., Coleman S., (2008) Modern-day perception of historic footwear and its lindd to preference, Journal of Applied Statistics, 32(2) pp 161-178, https://doi.org/10.1080/02664760701775498

Prithivirat, A., & Grzechnik, D. (2011). The Influence of Chocolate and Wafer on customers: An Application of Kansei Engineering. In *Institute of Technology*. Linköping University.

Schippert, O., Ottoson, T., Winlöf, R., & Oscar Lindskogen. (2017). *The Interface*.

Schütte, S. (2005). Engineering emotional values in product design-Kansei Engineering in development. In *Institution of Technology*. Linköping University.

Schütte, S. (2013). Evaluation of the affective coherence of the exterior and interior of chocolate snacks. *Food Quality and Preference, 29*(1). https://doi.org/10.1016/j.foodqual.2013.01.008

Schütte, S., Alikalfa, E., Schütte, R., & Eklund, J. (2006). Developing Software Tools for Kansei Engineering Processes: Kansei Engineering Software (KESo) and a Design Support System Based on Genetic Algorithm. *QMOD 2006*.

Schütte, S., & Almagro, L. M. (2013). *Development of an Affecitve Sensorial Analysis Method for Food Industry*.

Schütte, S., Eklund, J., Axelsson, J., & Nagamachi, M. (2004). Concepts, methods and tools in Kansei engineering. *Theoretical Issues in Ergonomics Science, 5*(3). https://doi.org/10.1080/1463922021000049980

Schütte, S., Marco-Almagro, L. (2013). *Development of an Affective Sensorial Analysis Method for Food Industry*.

Schütte, S., & Marco-Almagro, L. (2022). Linking the Kansei Food Model to the General Affective Engineering Model. *Internationasl Journal of Affective Engineering, 21*(3). https://doi.org/10.5057/ijae.IJAE-D-21-00024

Schütte, S., & Schütte, R. (2009). *Kansei Engineering Software (KESo)* (p. Data collection and evaluation software for Kansei). Linköping Institute of Technology. www.Kanseiengineering.net

Shimizu, Y., & Jindo, T. (1995). A fuzzy logic analysis method for evaluating human sensitivities. *International Journal of Industrial*

Ergonomics, *15*, 39–47.

Thorsson-Mendoza, C., Rappillard, M., Falcone, M., & Rubensson, J. (2021). Tech Toolbox for University Students. In *Student course project poster TMKT71*. Linköping University.

Tipping, M. E. (2001). Sparse Bayesian learning and the relevance vector machine. *Journal of Machine Learning Research*, *1*(Jun), 211–244.

Tomico, O., Mizutani, N., Levy, P., Yokoi, T., Cho, Y., Yamanaka, T., & others. (2008). Kansei physiological measurements and constructivist psychological explorations for approaching user subjective experience. *DS 48: Proceedings DESIGN 2008, the 10th International Design Conference, Dubrovnik, Croatia*, 529–536.

Tsuchiya, T., Ishihara, S., Matshbara, Y., Nishino, T., & Nagamachi, M. (1999). A Method for Learning Decision Tree using Genetic Algorthm and its Application to Kansei Engineering System. *IEEE SMC'99*.

Ueda, R., Araki, T., Sagara, Y., Ikeda, G., & Sano, C. (2008). Modified Food Kansei Model to Integrate Differences in Personal Attributes between In-house Expert Sensory Assessors and Consumer Panels. *Food Science and Technology Research*, *14*(5), 445–456. https://doi.org/10.3136/fstr.14.445

van Lottum, C., Pearce K., Coleman S., (2006). Features of Kansei Engineering Characterizing its Use in Two Studies: Men's Everyday Footware and Historic Footwear. Quality and Reliablity Engineering International 22(6) 629-650. https://doi.org/10.1002/qre.803

William AE, Nester CJ. Patient Perceptions of Stock Footwear Design Features. Prosthetics and Orthotics International. 2006;30(1): 61-71

PART 4

Exercises small group work and ideas

This section of the book provides practical exercises for small group work, using a fictitious development project for toothbrushes as an example. These exercises can be done either as self-study or group work, and a guide for teachers is included in the introduction. It is important to note that the methods and tools used in this context were chosen by the authors based on their feasibility and proven effectiveness.

Other methods may also be effective, and the choice here is not exclusive. Furthermore, it is important to recognize that after studying this chapter, the reader will have a basic understanding of the Kansei methodology, but will not become an expert. However, this knowledge will facilitate the reader's ability to comprehend and engage with expert material published in scientific contexts.

Chapter 14: Exercises and Teacher guidelines

Teacher guidelines

One of the main reasons for writing this book was the lack of teaching material for Kansei Engineering methodology. At Linköping University, Kansei Engineering is part of a course on affective aspects in product development with the undergraduate engineering programs.

Product development cannot be taught in a traditional classroom setting, but in order to spark creativity and problem solving ability it is necessary to allow for discussion, brainstorming and contemplation.

One teaching method that has been proven as an effective tool in teaching those topics is "Flipped Classroom methodology".

Flipped classroom is an instructional model that reverses the traditional method of teaching. In a flipped classroom, students are introduced to new content or concepts outside of class time, usually through video lectures, online modules, or reading assignments. They can access these materials at their own pace and convenience, often using digital technology.

Class time is then used for interactive activities, such as problem-solving, discussion, and collaborative projects. The teacher serves as a facilitator or coach, providing guidance and support as students work on applying and mastering the content. This approach allows students to engage in more active and personalized learning experiences and helps them develop higher-order thinking skills.

The flipped classroom model is based on the idea that learning is not limited to the classroom and that technology can facilitate more flexible and dynamic learning experiences. The flipped classroom can be used in various subjects and grade levels, and it has been shown to be effective in improving student engagement, motivation, and learning outcomes.

> This chapter includes some of the material used in the original course at Linköping. If you are a teacher you may use this material in this part of the book for free but make sure it is properly referenced in your and the students work.

Introduction to the exercises

The exercises in this chapter adhere to the general Kansei Engineering model proposed by Schütte et al., (2004), which is a compilation of methods typically employed in Kansei Engineering studies. See Figure 14.1. The model builds on over one hundred Kansei Engineering studies conducted mostly in Japan and Korea from 1970 to 2005, and subsequently refined using research data from a European context until 2020. Note that the model has been tested and adapted primarily in a European context. In Japan, where Kansei Engineering originated, it is also a philosophy for product development. Therefore, from this perspective, the Kansei Engineering model is somewhat restricted. Nevertheless, this presents an opportunity for practitioners to not only utilize the suggested tools for each step but also to introduce new tools and improve the functionality of the model.

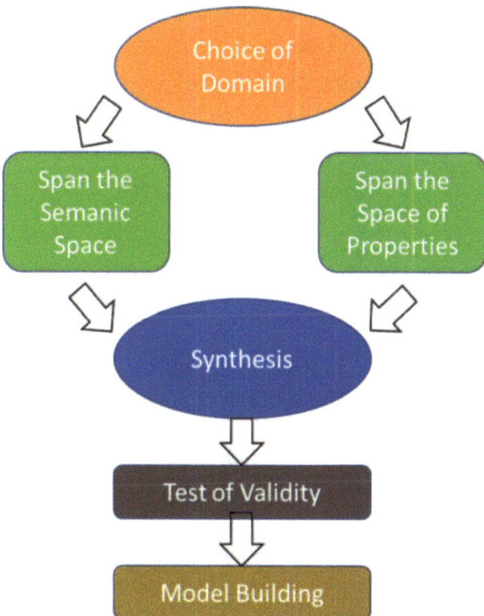

Figure 14.1: Universal Kansei Engineering model as proposed by Schütte et al., (2004).

In the upcoming exercises, we will walk you through a typical Kansei Engineering study using this model. We have designed the tasks to be straightforward and easy to comprehend. While the examples we provide may not necessarily be authentic or practical for product development purposes, they serve to illustrate the methodology in a clear and concise manner. It is essential to note that Kansei Engineering is a complex technology that necessitates experience and specialized expertise beyond what is covered in these exercises. Completing these tasks will not make you an expert in the field, but they will provide you with a foundation to comprehend more advanced concepts found in academic literature.

Discovering the feeling the customer wants

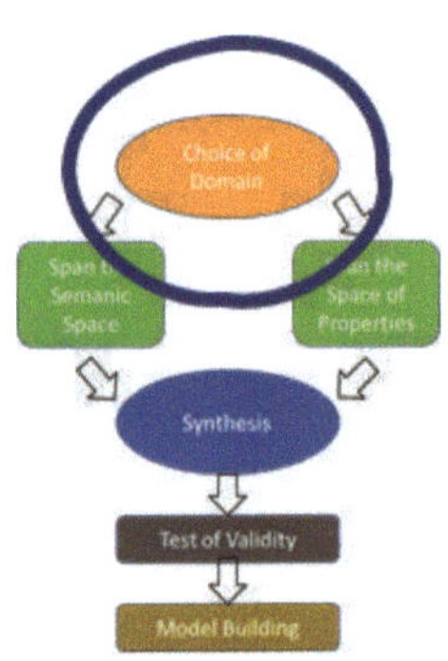

Users of physical products and services have a multitude of reasons for using them. While functionality is undoubtedly a crucial factor (e.g. a toothbrush is used for cleaning teeth), there is another dimension to customers' product choices that includes self-expression, prestige, social standing, and group belonging. Therefore, people often purchase products that enhance their self-image and allow them to convey their personality to others. A prime example of this is clothing; specific brands and styles are favoured by various groups to differentiate themselves from others.

In this exercise, you will gain a deeper understanding of the emotional needs of your intended target group. This will enable you to make informed choices for your product development, ensuring that your product resonates with your customers on a deeper level.

Goal:

The objective of this exercise is to identify everyday artifacts, products, and services used by the target audience to gain insights into their emotional needs, in addition to their functional needs. This information will help in designing a product or service that can meet the users' emotional needs and create a positive affective experience.

Preparation:

- Assemble a multifunctional group of 4-5 persons in a room.
- Instruct them to adopt a certain role: Management, marketing, production, etc.
- The product in question should be simple and familiar to everybody. You may choose any product or service. For the purpose of this exercise we have chosen a toothbrush.

You will need this material:

- Computer with internet access and big screen.
- Inspirational material (such as print material, sketches, ...)
- Collect a good variety of toothbrushes.
- Collect at least 5 different products that could be used in typical everyday situations by the target group.

Instructions:

1. Give a short introduction to the intended seminar and state the purpose.
2. Introduce the goal of the project which is to develop a new toothbrush for a predefined target group.
3. Present the everyday products to the group and tell them that this is the environment the new toothbrush will have to work in.

4. Let the participants experience the products for about 5 minutes.
5. Then let everyone describe verbally the impression they got from the target group with a focus on their role.
6. Now introduce the different toothbrushes you collected before.
7. In a short brainstorm session (max 5 minutes) let everybody come up with words that describe the affective appeal. Encourage them to use adjectives.
8. Let everyone get a good impression of the toothbrushes. Encourage them to touch and test.
9. Now let every member of the group choose one toothbrush that would fit the target group in accordance with their role.
10. Let everybody motivate their choice. Ask them specifically to point out what properties (colour, length, stiffness, etc.) they think have the most important affective appeal to the target group members.

Drawing conclusions for the next steps:

If possible record this session using camera and sound recorder. Also, you need to document, the statements from steps 5, 7 and 10. Verbal statements from step 7 will be needed for the Semantic Space in the next step and the toothbrushes selected as well as the properties mentioned in step 10 will feed into the creation of the Space of Properties.

Understanding the consumers affective needs

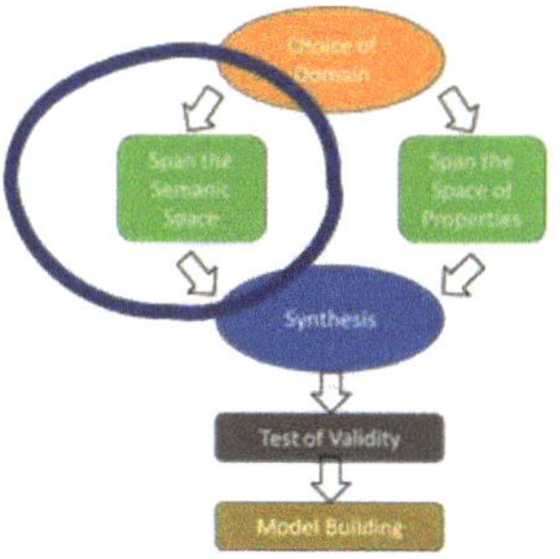

The general Kansei Engineering model (Figure 14.1) is a tool that describes the domain, defined in the previous step, from two distinct perspectives: The Semantic Space and the Space of Properties. In this step, we will focus on defining the former before moving on to the Space of Properties in the next step. These two steps can easily be carried out in parallel or within the same seminar during a real product development process.

As previously mentioned, the Semantic Space originates from the work of Osgood (1969), who developed a methodology called Semantic Differentials to connect verbal terms with objects. This methodology was later used to uncover hidden patterns in word meanings. For example, when collecting words commonly used to describe food, words such as "tasty," "good," "nourishing," "fancy," "hungry," and so on would be gathered. Using Principle Components Analysis (PCA), distinct dimensions would emerge, including taste-related terms, physical effects of food, and prestige aspects.

Pattern finding can be accomplished using statistical tools like PCA, but manual sorting methods can also produce good results while using less data and time. The Semantic Space used in the context of Kansei Engineering employs similar tools. It groups words that have a positive affective impact on the target audience and visualizes the relationship between how people perceive certain products from an affective perspective.

Goal:

The objective of this step is twofold: Firstly, to define the Semantic Space and secondly, to identify the underlying structure of how users perceive products from an affective standpoint. Additionally, the aim is to determine the factors that make products appealing to users.

Preparation:

- Assemble a group of "lead users" of the product in question. These users need to belong to the target group defined in the previous step. In the case of this exercise toothbrushes will be used as a sample product. The users should have good knowledge and preferably personal experience with the usage of the product. It would also be good if some of them have knowledge about the technical details of the product. Since toothbrushes are a very common product, basically everybody can be considered a lead user. If you do this exercise in class, the whole group can be engaged.
- Make sure that the participants are talkative and can freely express their opinions.

You will need this material:

- Sticky notes
- A white board or similar space for setting up the sticky notes

Instructions:

1. Begin by introducing the task to the group. In this case, let them know that they will be verbally describing toothbrushes.

2. Display all the collected toothbrushes and highlight the ones that have been identified as particularly suitable for the target group.

3. Give the group time to familiarize themselves with the toothbrushes. Encourage them to use their senses as much as possible, as before.

4. Distribute sticky notes to each group member. Ideally, each member should receive sticky notes of a different color.

Figure 14.2: Sticky notes in a different colour should be provided for each participant.

5. Instruct the group to brainstorm as many descriptive words as possible, writing each one on a separate sticky note. Emphasize that they should focus on feelings and subjective perceptions rather than technical or physical features. Ideally, the words they generate will be adjectives. This step should be carried out individually, and silence is allowed. Continue until no new words appear, which should take between 5 and 15 minutes depending on the product.

6. Allow the participants to take a break while you put up the sticky notes on a whiteboard or similar surface, removing any duplicates. See Figure 14.2.

7. Once the group has reconvened, give them a few minutes to look over all the words. Discussion is allowed during this time.

8. Instruct the participants to cluster the words into groups that have a meaning-wise correlation. They should do this without talking to each other. If there is disagreement about particular words, ask the participants to set those words aside for further discussion later on. See Figure 14.3.

Figure 14.3: Using Card System for defining the Semantic Space.

9. Allow time for the groups to be created and finalized.
10. Take up the sticky notes with words that were set aside and discuss them with the group.
11. Finally, document the clusters that were created.

Drawing conclusions for the next steps:

Kansei words represent the various affective dimensions or factors that users and customers consider when interacting with a product or service. Typically, there are several key affective dimensions that are most important to users, and designers can be more mindful of these dimensions when developing product concepts.

For instance, when it comes to apartment design, some important affective dimensions may include the concrete factor (such as practicality, good planning, and spaciousness), the exclusiveness factor (featuring words like fancy, elegant, and luxurious), the ambient factor (relating to comfort, snugness, and coziness), and the lighting and harmony factor (describing lightness and harmony).

In the case of toothbrushes, experience factors such as performance, product style, convenience and hygiene may be particularly relevant (Table 14.1). By paying close attention to these key affective dimensions, product and service designers can create products that are more likely to resonate with their target audience and meet their needs and expectations.

- In order to continue to the next step, select one word from each cluster in order to represent the whole cluster. Document those clusters and corresponding words. If you are dealing with big clusters you may also use 2 or more words to represent those. In rare cases it might also be opportune to find a new Kansei word representing the cluster in question.

Table 14.1: Example of Semantic Space for toothbrushes. Bold text describes the semantic dimension and the red marked words represent the meaning of its cluster.

1. **Performance and Effectiveness**: *Sensitive, Gentle, Soft, Bristly, Ergonomic, Oral-care*
2. **Style and Design**: *Stylish, Colorful, High-tech, Advanced, Basic*
3. **Durability and Price**: *Durable, Cheap, Affordable, Reusable*
4. **Convenience and Portability**: *Practical, Convenient, Portable*
5. **Hygiene**: *Hygienic, Clean, Fresh*

The collection and aggregation of Kansei words and clustering of the same is traditionally done in groups. Due to personal interaction this seems to date to be the most accurate way to do it. However, there are also studies done where this is done online (Valverde & Schütte, 2022). Figure 14.7 demonstrates an example of how this could be done.

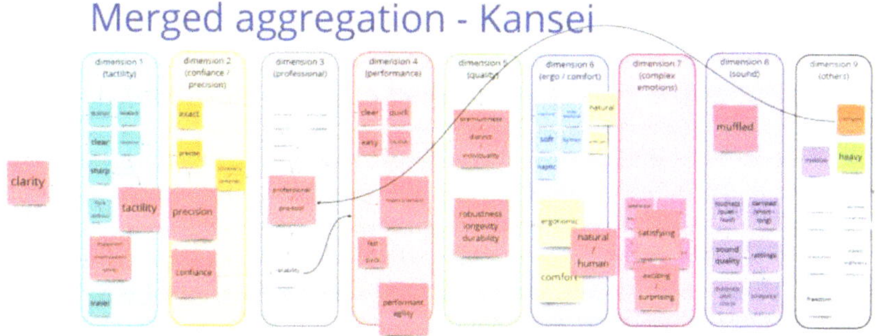

Figure 14.7: Kansei Word aggregation carried out on an electronic white board.

Select affective product items and create corresponding product samples.

Understanding the products delighters

The Space of Properties in Kansei Engineering refers to the set of physical and sensory properties of a product or service that are relevant to the affective experiences or emotions that users associate with it. In other words, it is the multidimensional space of sensory and affective attributes that can be used to describe and quantify the subjective feelings and emotions that users have when interacting with a product or service.

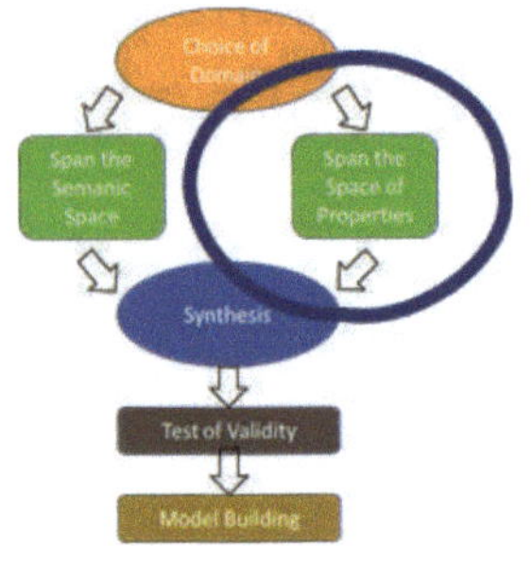

The Space of Properties can include various dimensions, such as color, shape, texture, sound, smell, taste, temperature, and weight, among others. These dimensions can be further divided into sub-dimensions or properties that are relevant to the specific product or service being studied.

Goal:

The aim of this step is to identify the physical factors that have the most significant impact on the user's emotional response when using toothbrushes. By understanding which key physical factors are most relevant to users, designers can create mock-up concepts that combine these factors in different ways to be tested against customer preferences. This process helps prioritize the key factors to determine which are the most important and which can be excluded from further evaluation. By doing so, the data is simplified and easier to analyze, which can inform the development of toothbrushes that better meet users' emotional needs and preferences.

Preparation:

- Ideally, this step should be conducted in the same seminar as the previous step of establishing the Semantic Space. If conducted separately, it is crucial to assemble lead users who have in-depth knowledge of the product and belong to the target domain identified in the first step.
- Collect additional material that represents a variety of possible key properties such as different materials, color samples, size measurements, and other relevant factors. This helps designers explore the physical key factors that have the most affective impact on the user of the product.
- Gather material that depicts futuristic or unusual concepts for the product group in question. This can help to generate innovative ideas and stimulate creative thinking, leading to designs that meet customers' needs and preferences.
- By bringing together lead users and relevant materials, designers can create mock-up concepts that can be tested against customer preferences. This process helps to identify which key factors are most important to users and prioritize them accordingly, enabling designers to exclude factors with negligible impact, making the data easier to analyze.

You will need this material:

- The collection of toothbrushes used in the previous step.
- Sticky notes.
- White board or similar for setting up the sticky notes.
- Computer with electronic spread sheet software.

Instructions:

1. Introduce the participants to the task. Ideally 5 persons per group can fulfil the task in the best way. Allocate similar roles to each participant as in the previous step (Management, marketing, production, etc.) Point out that in contrast to the previous step of selecting the Semantic Space they now are supposed to focus on physical features of the toothbrushes.
2. Ask the participants to familiarise themselves with the product samples, concepts and additional material provided. Encourage them to think aloud or talk to each other. Allow 5-10 minutes for this step. When the discussion among them dies out this is typically a sign to proceed.
3. Now provide sticky notes to them. Each participant should receive her own colour in order to visualize their contribution to the total resulting data.
4. Ask the participants to brain storm features, assets and properties that they personally perceive as important for shaping a good affective impact on future users of the product. Those can be colour, size, material, modes of function, layout, etc.
5. Collect all sticky notes on the white board and eliminate duplicates.
6. Let all participants familiarize themselves with the results.
7. Then tell the participant that they have a total of 100 points to spread out between the different features identified. The more important they perceive the particular feature for the potential affective impact, the more points they should allocate.
8. Document the results.

Drawing conclusions for the next steps:

In order to analyse and assemble the data in a sensible way a method called PARETO Chart (e.g. Bergman & Klefsjö, 2002) will be used. At this point, you have assigned weights to each of the features. Using spreadsheet software, you can assemble a table that shows the feature and the number of votes it received. Additionally, you can calculate the relative percentage of votes for each feature, as well as the cumulative percentage.

illustrates this calculation, using the toothbrush study as an example also used in the earlier steps. The table is ordered according to the number of votes each feature received, and can be interpreted as a prioritization list. The more votes a feature received, the more important it is to the participants (i.e. users) on an affective level.

For instance, considering the second row of the table, we see that the means of propulsion for the toothbrush was mentioned 195 times, which represents 32.5% of the total votes. When combined with the colour feature depicted in the row above, the cumulative percentage is 79.2%. As a result, these two features can be regarded as the two most important features to a toothbrush designer.

Table 14.2: Summary table from the data collection for the Space of Properties.

Column1	Mentionings	Percentage	Culmulative Percentage
Colour	280	46,66666667	46,66666667
Propulsion	195	32,5	79,16666667
Material	37	6,166666667	85,33333333
Stiffness	25	4,166666667	89,5
Form	25	4,166666667	93,66666667
Length	12	2	95,66666667
Widht	9	1,5	97,16666667
Bristles	8	1,333333333	98,5
Shaft width	7	1,166666667	99,66666667
Grip	2	0,333333333	100
Total	600		

Figure 14.5 presents this data visually. The blue bars represent the number of votes received, while the grey line graph displays the cumulative importance of the features. This graph makes it clear that the Kansei study cannot consider all of the features mentioned - in theory, there could be thousands of them. However, some features, such as "grip" or "shaft width", were deemed unimportant to users from an affective perspective, as evidenced by their low number of votes.

The next step is to determine how many and which features to select for further consideration. There is no exact cut-off point defined, but a general rule of thumb is the eighty-twenty rule, which is also illustrated in ***Figure 14.5: Example of a Pareto Chart presenting 80/20 rule.***

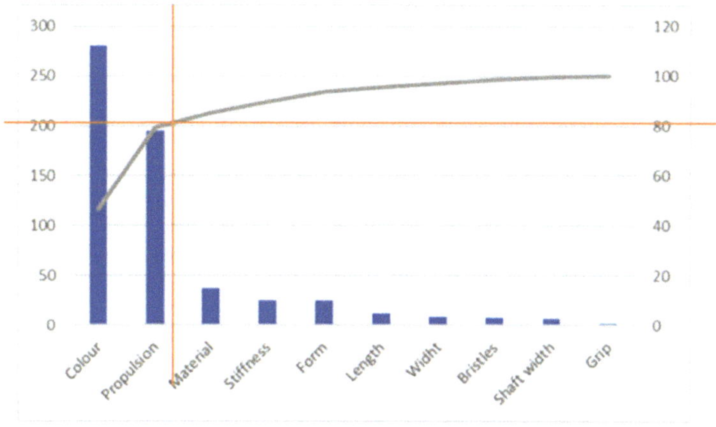

Figure 14.5: Example of a Pareto Chart presenting 80/20 rule.

As can be observed, the "colour" and "means of propulsion" features comprise approximately 80% of the cumulative percentage, even though they account for only about 20% of the total number of evaluated features. This is the basis of the 80/20 rule. Furthermore, it is evident that these two features are the most important ones from an affective standpoint, as they completely dominate the users' voting behavior.

It is important to note that the toothbrush example was selected deliberately to demonstrate this point. In other studies involving more complex products, there may be additional features that qualify within the top 20%. However, for the purposes of this discussion, we will focus solely on the "colour" and "means of propulsion" features.

In the context of Kansei Engineering research, these selected product features are referred to as "items." In order to identify the most suitable "categories" for each item, there are several methods that can be employed. The most common approach is to involve company or client personnel in the selection process, taking into account factors such as producibility, input from the designer, benchmarking, and so on. It is crucial to ensure that all selected categories are feasible and achievable in the subsequent stages of product realization. In the toothbrush study, two categories were chosen for each item as an example. Table 14.3 illustrates the categories the team decided on.

Table 14.3: Categories found for the items identified.

Colour	Propulsion
white	manual
other	motor

In the subsequent step of the synthesis phase, it is necessary to create sample products that can be evaluated by future lead users. These products must be presented to users in a way that engages their senses appropriately, which may involve different methods of representation depending on the particular product features being evaluated. For instance, an overall car design could be evaluated through a rendering, while a new chocolate filling will require tasting and new material choices for a car dashboard requires touch. The mock-ups presented in the study do not necessarily have to be production models, and in fact, they should include as few additional features as possible beyond those chosen in table 14.3. This reduces nuisance factors affecting the later validating behaviour negatively. It can be advantageous for the samples to be somewhat more "extreme" than the production models, as this will make user reactions clearer and facilitate easier conclusions. Table 14.4. illustrates all possible combinations for the toothbrushes.

Table 14.4: Product samples necessary for evaluation (full factor experiment).

Name	Colour	Propuls
Sample 1	white	manual
Sample 2	other	manual
Sample 3	white	motor
Sample 4	other	motor

The more categories chosen in the previous steps, the more samples have to be found, prototyped and finally evaluated by the lead customers in the next step. It is obvious that this is limited due to factors such as time and fatigue. If you wish to use more categories than the users can rate, you would need to employ a so called "reduced factorial experiment". Here you need to use advanced statistical tools. Some commercially available software has that functionality.

In the context of the toothbrush study, only images were used to present the sample products to the users. This approach has limitations as it does not allow for the evaluation of factors such as weight, material, and flexibility, which might also be important for a thorough assessment. In a real study, these factors should not be neglected. However, for the sake of simplicity in the book's context, only visual cues were used.

Figure 14.6 shows a possible sample for the study. Those samples have been generated by a AI software. In comparison to a manual sketching technique this a faster alternative and creates better renderings in shorter time but on the other hand it cannot creatively produce a new design - yet.

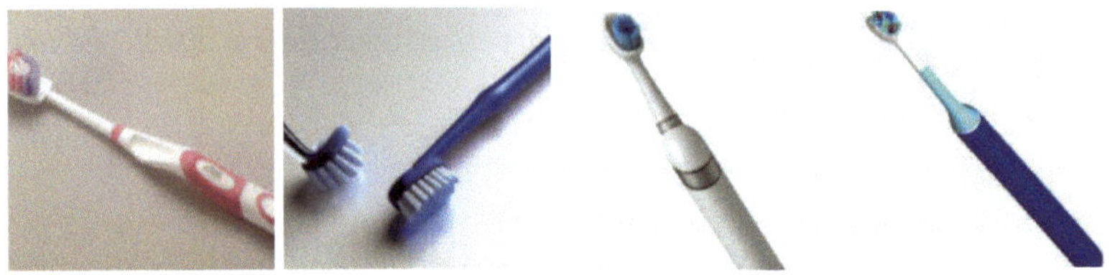

Figure 14.6: Possible product samples generated using an AI software (samples 1- 4 from left to right).

Linking the Emotions to product features

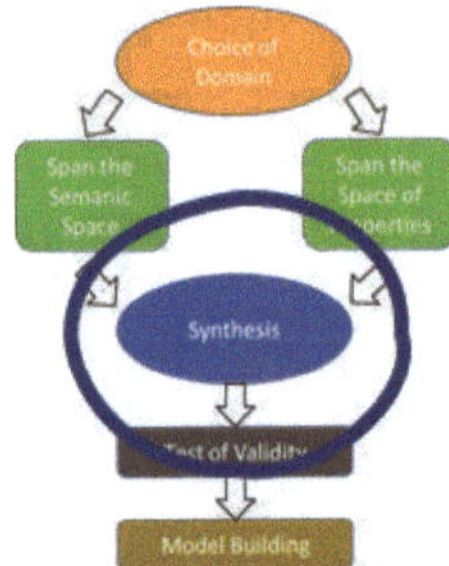

The synthesis phase is unique to Kansei Engineering methodology. It combines the Semantic Space and the Space of Properties. This phase establishes connections between the various Kansei Words and the guiding properties, providing essential assistance to product and service designers while making design decisions. It is important to note that Kansei Engineering has different methods available for achieving this linking. Those methods can be broadly categorized into mathematical and sorting methods. However, in order to make this part comprehensive, this book only focuses on a selected number of methods, such as Quantification Theory Type 1 and manual evaluation, with the use of Kansei Engineering Software for data collection and evaluation.

Before proceeding with data analysis, the data must first be collected. Part 1 of this book thoroughly explains various types of data collection scales, their advantages, and drawbacks. In this particular study, Visual Analogue Scales (VAS) were used. ***Figure 14.7*** demonstrates the data collection site for the toothbrush example in the previously mentioned Kansei Engineering Software tool, KESo. Participants were initially asked about relevant demographic information such as gender, age, and experience with the products or services in question. Then, the collection software presented each product sample as selected in the definition of the Space of Properties above a list of VAS rating scales featuring the Kansei words as defined in the Semantic Space.

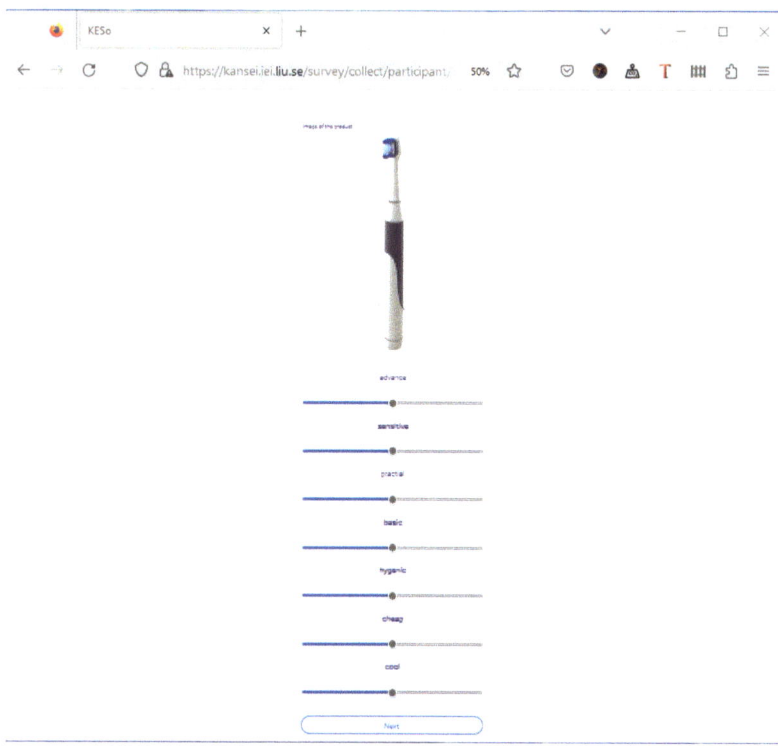

Figure 14.7: Data collection site in Kansei Engineering Software.

This results in a three-dimensional data set as displayed in Figure 14.8. For each of the study participants their ratings for each set of Kansei Words for each product sample is stored. This poses a problem for the visualization of the data and possible further mathematical treatment. Hence, this three-dimensional matrix is transferred into two possible two-dimensional matrices by averaging either the participants' ratings or averaging the product samples' ratings.

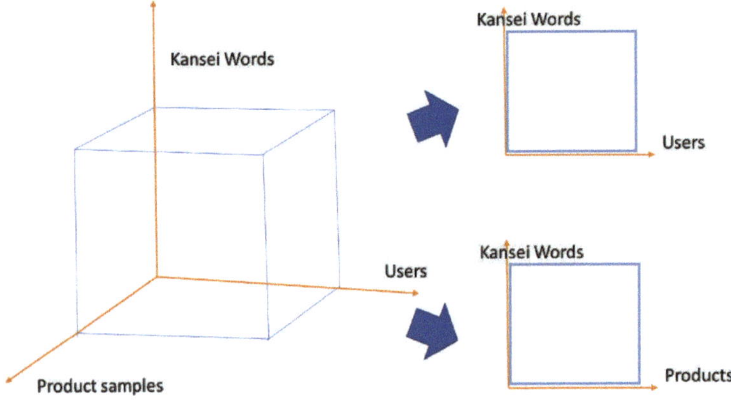

Figure 14.8: 3D structure of Kansei Engineering data reduction to 2D data. (upper right: average over products; lower right: average over users).

Goal:

The aim of this part is to identify links between the Semantic Space (users' affective needs) and the Space of Properties (product and service designers' way to evoke a desired user affect). This will happen in three steps:

1. Design of user questionnaire
2. Data collection
3. Data evaluation

Kansei Engineering Software has been presented in Part 1. This software can do all of those steps in a mostly automatized way. In this exercise the steps carried out by the software are done manually and in a simplified way to properly understand the process.

Preparation:

- Make sure you have the Kansei Words at hand.
- Make sure you have the Items (Product features), the selected categories and the chosen samples from the previous step at hand.
- Identify, select and book time with potential participants of your study.

You will need this material:

- Spread sheet software.
- Printer if you chose to collect customer data manually on paper (alternatively you can use commercially available data collection software including Kansei Engineering Software).
- Statistical Software package (if you wish to execute advanced data analysis).

Instructions for data collection:

1. Design a data collection sheet as presented in ***Figure 14.7***. Use a VAS scale for each Kansei Word and present your product on top of them. Use separate sheets for each product or service.
2. Collect data with your pre booked participants.

3. Once all data is in, assemble your data file. As explained above it will comprise a three - dimensional matrix. This matrix has to be transformed into two two- dimensional matrices (compare ***Figure 14.8***). You can use the example file from the toothbrush study as an inspiration. Please note that these are for demonstration purposes only. Do not assume real conclusions for the development of toothbrushes.

Instructions for data evaluation:

1. Load the two-dimensional data matrix averaged over participants as in Figure 14.9.

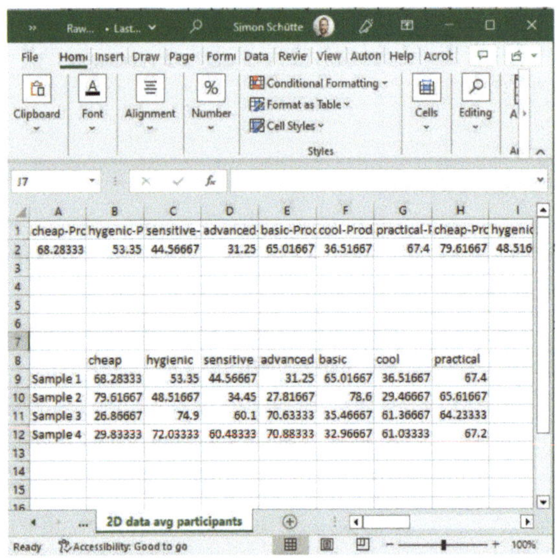

	A	B	C	D	E	F	G	H	I
1	cheap-Prc	hygenic-P	sensitive-	advanced-	basic-Proc	cool-Prod	practical-I	cheap-Prc	hygenic
2	68.28333	53.35	44.56667	31.25	65.01667	36.51667	67.4	79.61667	48.516
8		cheap	hygienic	sensitive	advanced	basic	cool	practical	
9	Sample 1	68.28333	53.35	44.56667	31.25	65.01667	36.51667	67.4	
10	Sample 2	79.61667	48.51667	34.45	27.81667	78.6	29.46667	65.61667	
11	Sample 3	26.86667	74.9	60.1	70.63333	35.46667	61.36667	64.23333	
12	Sample 4	29.83333	72.03333	60.48333	70.88333	32.96667	61.03333	67.2	

Figure 14.9: Two-dimensional data matrix averaged over participants.

2. Create a radar chart in the table as in Figure 14.10.

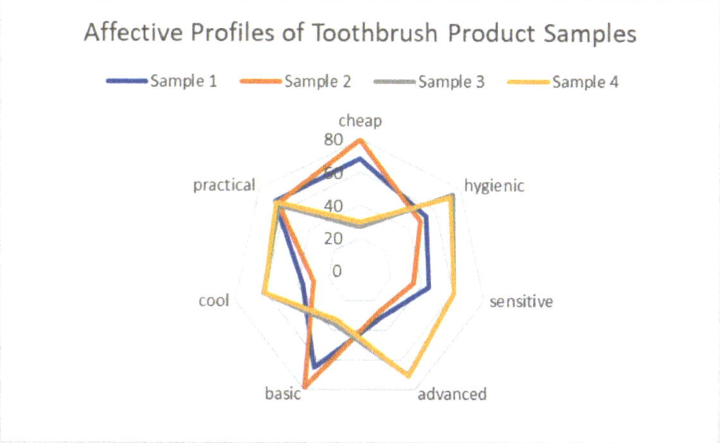

Affective Profiles of Toothbrush Product Samples

Figure 14.10: Radar chart illustrating affective profiles for product samples.

The radar chart is the easiest way to examine your data. It presents the affective profile of the examined products. If you as a designer already have a preference for the future product's intended fingerprint you can already see which of the samples are closest to it.

A toothbrush designer might prioritise "hygienic and sensitive" for the new product. In the radar chart in Figure 14.10 from the toothbrush study she can see that samples 3 and 4 rank high for those items. On closer inspection she will also find out that the commonality they share is the fact that both have a motor. If a motor is not an option then product sample 1 may be an alternative; it is white and manual and rates higher than its blue counterpart on "hygienic" and "sensitive".

Using Quantification Theory Type 1

Using Quantification Theory Type 1 also reveals a similar conclusion: the impression of "hygienic" is achieved foremost by the usage of a motor for the propulsion, secondarily by the colour white (Table 14.5).

Table 14.5: Excerpt from evaluation using Quantification Theory Type 1.

hygienic	0.994	colour	0.500	White	2.389
				Other	-2.389
		propulsion	0.025	Manual	-14.643
				Motor	14.643

Using KESo Software

The functionality of KESo software is demonstrated in a study that evaluates toothbrushes by real consumers. This study serves as an example of how the software can be used. Toothbrushes were chosen for their simplicity, few variations, and cultural neutrality. They represent a group of products which are familiar to many people and have a stable design that has not changed for a long time. Additionally, toothbrushes are personal property and are used in close proximity, resulting in a good Kansei and making toothbrushes particularly suitable for Kansei Engineering evaluations, as they are experienced with all senses.

The purpose of the study is to showcase the functionality of the KESo software, which can be used to analyze and understand consumers' emotional responses to products. The evaluation of toothbrushes by consumers provides valuable insights into how users perceive and feel about the product, which can inform product design and marketing strategies. This study is a demonstration of the potential uses of the KESo software in understanding consumer preferences and emotions towards products.

By scanning the QR-link here you will be guided to the original data collection site. The whole study takes about 5 minutes to carry out and can serve as an experience of how Kansei Engineering data collection can be done.

The **data collection** method used in this study utilizes visual analogue scales (VAS), which offer several benefits over traditional Likert scales occasionally also used for Kansei data collection. VAS scales allow participants to provide ratings without being constrained by predefined categories, which improves the quality of data collected. Additionally, using VAS scales helps to ensure that the data collection process is efficient and does not waste time by requiring participants to "think about" their rating. This is especially important in studies where fast ratings are desired because they are more likely to reflect participants' initial emotional responses. In contrast, slow ratings can encourage rational cognitive decision-making, which can bias the results towards less emotional responses. By using VAS scales, this study can obtain accurate, efficient, and emotionally-based ratings from participants. Compare (Schütte, 2005; Marco-Almagro, 2011a).

Furthermore, it is worth noting that the scales used in this study do not have any descriptors at their extremes. While it is possible to include descriptors through software settings, doing so can lead to a more cluttered data collection form and subsequent slower ratings by the participants. This is clearly undesirably as discussed above. Furthermore, in most cultural contexts, people tend to agree that higher ratings are associated with being further to the right on the scale. However, this principle only works if all the chosen words have positive connotations, such as "good" instead of "bad." If the words have negative or mixed connotations, the user may need to convert the scale, which takes more time and can result in the aforementioned drawbacks. It is important to note that there is currently no consensus among researchers on this matter, and as a result, the software used in this study allows for optional nomenclature to be added to the ends of the scales. Schütte (2005) explores the differences, advantages, and drawbacks of various approaches to scale design in his work.

The KESo software offers a range of evaluation tools for data analysis. However, ensuring the data's reliability and quality is crucial, which requires proper screening of respondents before collecting data. By stratifying the data based on screening questions, researchers can identify possible differences within sub-groups, and by downloading the data in .xml format, they can conduct further analyses using other software.

For **data evaluation** the data collected is then fed into a three-dimensional databank from which all further evaluations are done automatically. KESo software can perform various evaluations, including Quantification Theory, Ordinal logistic model, and mixed model ordinal logistic model. However, for reliable results, it is essential that the data is normally distributed and free from bias. To ensure this, it is necessary to screen the respondents' suitability and background properly beforehand.

KESo provides an option to stratify the gathered data based on screening questions, such as age, gender, income, and familiarity with the product or service in question etc. This allows the study conductor to compare data drawn from sub-groups of participants and spot possible differences within the previously determined stratification criteria, such as different age groups. These can then be acted upon in order to make the product or service desirable for all sub-groups.

For a more in-depth examination of the data quality, the data can be downloaded as an .xml file, which can be easily imported into other commercially available data evaluation packages such as SPSS. This enables researchers to conduct more comprehensive analyses of the data.

As its main output, KESo delivers tables derived from the latter mentioned tools. Table 14.6 shows results for the study on toothbrushes.

Table 14.6: KESo output table for selected Kansei words.

Kansei word	**MCC^2**	**Item**	**PCC**	**Category name**	**Category score**
practical	0.531	colour	-0.445	White	-0.071
				Other	0.071
		propulsion	-0.404	Manual	-1.700
				Motor	1.700
basic	0.965	colour	-0.239	White	-3.143
				Other	3.143
		propulsion	0.909	Manual	23.200
				Motor	-23.200
advanced	0.997	colour	0.465	White	0.757
				Other	-0.757
		propulsion	0.252	Manual	-22.700
				Motor	22.700
hygienic	0.994	colour	0.500	White	2.389
				Other	-2.389
		propulsion	0.025	Manual	-14.643
				Motor	14.643

As a first glance ***Table*** shows the evolution of the original study data on toothbrushes as it was presented above. The evaluation method Quantification Theory Type 1 was chosen but the results should lead to equal conclusions in the case of the other evaluation tools. The researchers in this study have chosen the colour (white or other colour) and the propulsion of the toothbrush (manual or with motor) as the two most important items influencing the affective user impression of toothbrushes.

The data can be interpreted following three steps.

1. Make sure that: MCC^2 > 0.7. The square of the Multiple Correlation Coefficient (MCC) must be above 70%. This measure indicates to what extend the linear model created by the software fits the data. It is generally accepted in Kansei Engineering that 70% is sufficient to draw valid conclusions.

2. Inspect PCC. The Partial Correlation Coefficient (PCC) gives an indication to what extent a particular item influences the subjective impression. The higher the value, the more important the item is to the user.

3. Determine the CS. The Category Score (CS) tells which combination of physical attributes is best suited in order to evoke a certain Kansei/ Emotion.

Following these rules the following findings can be derived:

- Since the MCC^2 for the Kansei Word "practical" is below 0.7 the data derived cannot be used at all (grey values).

- As for the Kansei word "basic"; the MCC^2 allows interpretation of the data. It shows that propulsion influences the customer rating more strongly than the colour ($PCC_{Propulsion}$>PCC_{Colour}). Also "manual propulsion" and "other colour" yield the highest CS for each item. This means that the affective feeling of "basic" is driven by those product features.

- The Kansei Words "advanced" and "hygienic" show similar results. Both ratings are driven by colour before propulsion and in both cases the combination of white and motor-driven toothbrushes is optimal.

The study found that people perceive toothbrushes rated as "advanced" and "basic" as opposite based on different combinations of features. The "hygienic" rating was linked to motor-driven toothbrushes, which makes sense objectively. The color white was also associated with hygiene since many professional dentists use white-colored equipment.

Based on the study, the ideal toothbrush would be white and motor-driven. However, if the manufacturer needs to develop an affordable product without a motor, they can still achieve a "basic" and "hygienic" combination. The study shows that the "basic" rating is mostly driven by the toothbrush's propulsion, while the "hygienic" rating is driven by its colour. Therefore, a good trade-off would be a manual toothbrush that is white in colour.

Chapter 15: Examples of group work

Linköping University regularly offers an undergraduate course entitled "Affective Product Development," where students collaborate with industrial partners to enhance the emotional appeal of one of their products. The objective is to add another dimension to the product by increasing its affective impact, attracting users, or making it more desirable, or all three.

The course comprises multiple parts, including lectures to grasp the theoretical concepts, seminars for hands-on experience with the tools used in Kansei Engineering, and a project work conducted in small groups under the supervision of experienced Kansei Engineering experts. One of the course's prerequisites is for each group to publish a poster outlining their unique problem provided by an organization or company. The groups present their work in a final presentation and conclude with an individual oral or written exam.

This chapter showcases several posters that cover a wide range of product types, providing inspiration for practitioners, teachers, and students alike. Kansei Engineering is a complex methodology that is difficult to convey to individuals outside the field. Therefore, finding simple ways to present complex information is crucial to Kansei Engineers. These posters represent efforts to achieve this goal.

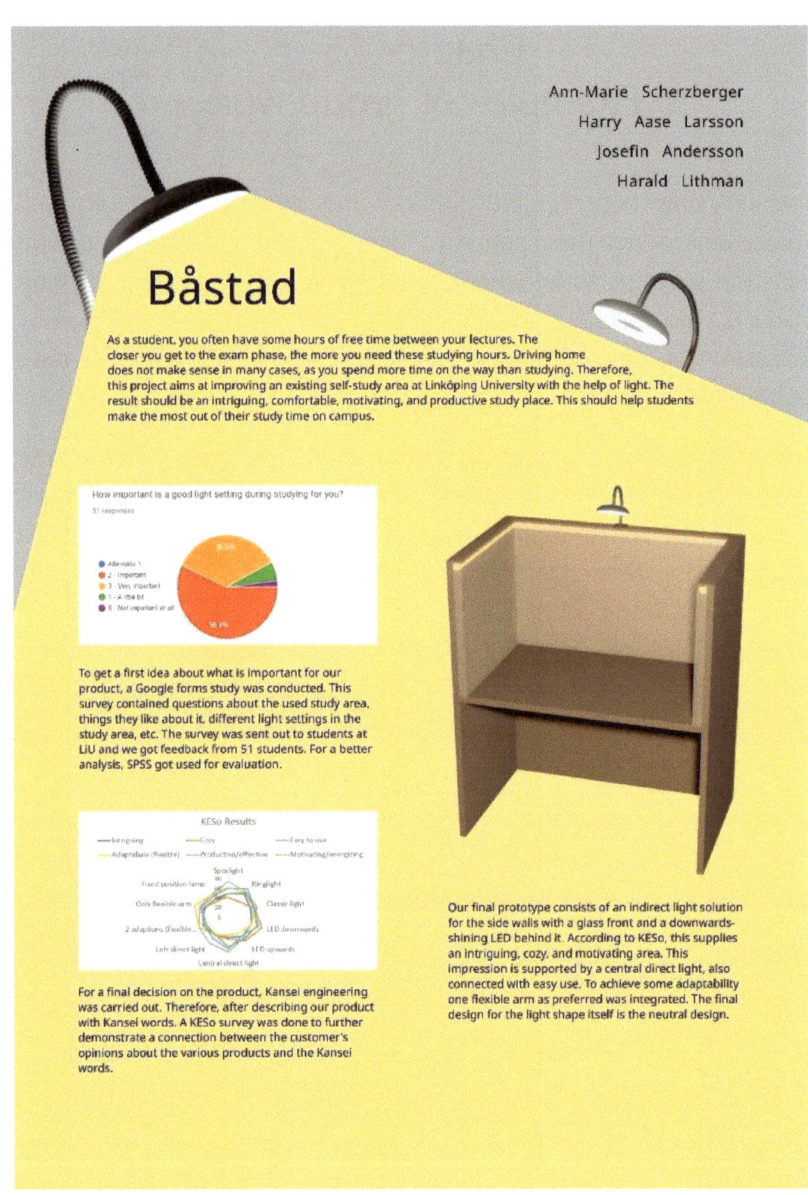

Figure 15.2: Student course poster "Båstad" (Scherzberger et al., 2022).

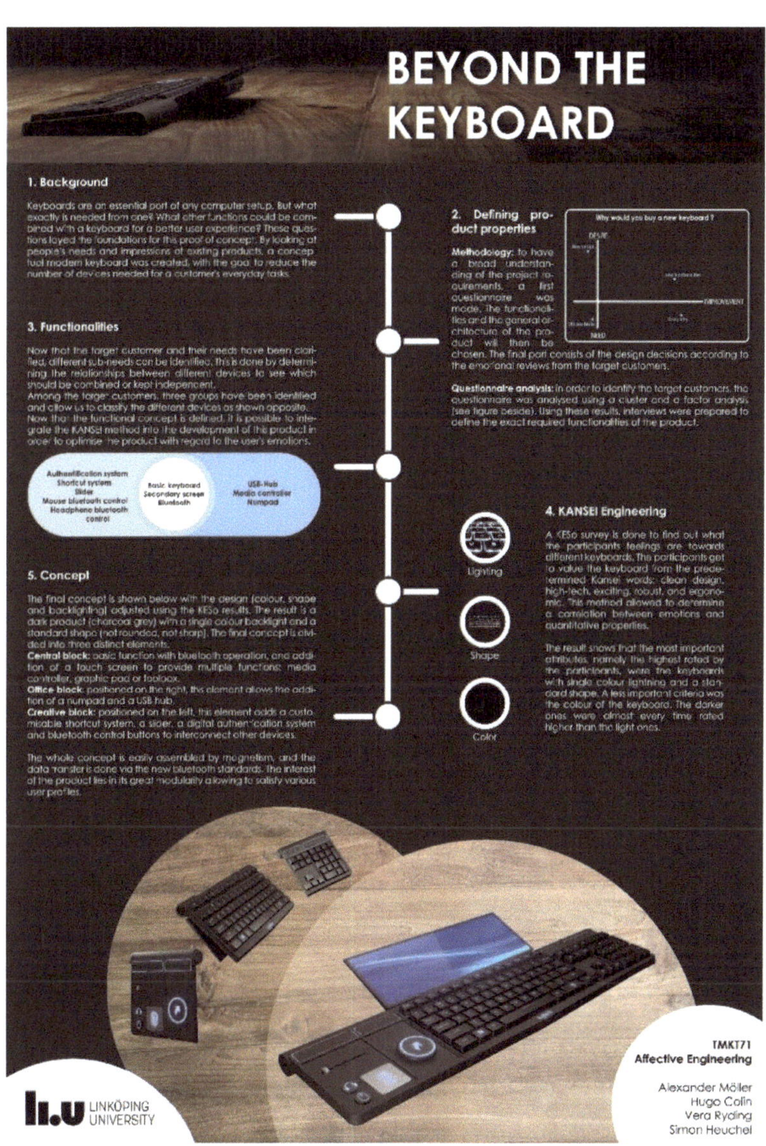

Figure 15.3: Student course poster "Beyond the keyboard" (Möller et al., 2022).

References for Part 4

Augustini, J., Jean, J., Jonsson, A., & Lefebvre, A. (2022). *Tear Tag.*

Bergman, B., & Klefsjö, B. (2002). *Kvalitet i alla led*. Studentlitteratur.

Eriksson, L., Eliyo, D., & Franzén, V. (2017). *Developing a Steering Wheel for an Autonomous Car.*

Falk, J., Fredriksson, T., Wallin, R., & Nguyen, N.-A. (2016). *Developing a Box Design for Ice Cream.*

Lundblad, F., Raujol, B., Rapp, J., & Wester, J. (2022). MX Master Hologram Edition. In *Student course project poster TMKT71*. Linköping University.

Lundvik, A., Quere, T. L., Mir, P., & Welander, T. (2021). *Headphones for kids age 5-6.*

Marco-Almagro, L. (2011). *Statisical Methods in Kansei Engineering Studies*. UPC Barcelona Tech.

Mensah, A., Schmeling, E., Gerklev, L., & Svenonius, I. (2017). *Private Space.*

Möller, A., Colin, H., Ryding, V., & Heuchel, S. (2022). *Beyond the keyboard.*

Osgood, C. E. (1969). The Nature and Measurement of Meaning. In C. E. Osgood & J. G. Snider (Eds.), *Semantic Differential Technique - a Source Book* (pp. 3–41). Aldine publishing company.

Scherzberger, A.-M., Asse-Larsson, H., Andersson, J., & Lithman, H. (2022). *Båstad.*

Schippert, O., Ottoson, T., Winlöf, R., & Oscar Lindskogen. (2017). *The Interface.*

Schütte, S. (2005). Engineering emotional values in product design- Kansei Engineering in development. In *Institution of Technology*. Linköping University.

Schütte, S., Eklund, J., Axelsson, J., & Nagamachi, M. (2004). Concepts, methods and tools in kansei engineering. *Theoretical Issues in Ergonomics Science*, *5*(3). https://doi.org/10.1080/1463922021000049980

Thorsson-Mendoza, C., Rappillard, M., Falcone, M., & Rubensson, J. (2021). Tech Toolbox for University Students. In *Student course project poster TMKT71*.

Linköping University.

Valverde, N., & Schütte, S. (2022). ACHIEVING CUSTOMER ACCEPTANCE OF NOVEL PRODUCT FEATURES BY OFFERING CUSTOMER DELIGHT. *9th International Conference on Kansei Engineering- KEER 2022,*.

Wilhelmsson, M., Sörberg, S., Brännström, O., & Martinsson, N. (2016). *Packaging for an exclusive watch*. Linköping University.

Willén, M., Fagergren, M., Jonsson, W., & Mittaine, M. (2021). *Children's Headphones*.